John Richard Green, Alice Stopford Green

A short Geography of the British Islands

John Richard Green, Alice Stopford Green

A short Geography of the British Islands

ISBN/EAN: 9783744729710

Printed in Europe, USA, Canada, Australia, Japan

Cover: Foto ©berggeist007 / pixelio.de

More available books at **www.hansebooks.com**

A SHORT GEOGRAPHY

OF THE

BRITISH ISLANDS.

BY

JOHN RICHARD GREEN, M.A., LL.D.,

HONORARY FELLOW OF JESUS' COLLEGE, OXFORD,

AND

ALICE STOPFORD GREEN.

WITH MAPS.

London:

MACMILLAN AND CO.

1879.

INTRODUCTION.

No drearier task can be set for the worst of criminals than that of studying a set of geographical textbooks such as the children in our schools are doomed to use. Pages of "tables," "tables" of heights and "tables" of areas, "tables" of mountains and "tables" of table-lands, "tables" of numerals which look like arithmetical problems, but are really statements of population: these, arranged in an alphabetical order or disorder, form the only breaks in a chaotic mass of what are amusingly styled "geographical facts," which turn out simply to be names, names of rivers and names of hills, names of counties and names of towns, a mass barely brought into grammatical shape by the needful verbs and substantives, and dotted over with isolated phrases about mining here and cotton-spinning there, which pass for Industrial Geography.

Books such as these, if books they must be called, are simply appeals to the memory; they are hand-books of mnemonics, but they are in no sense hand-books of Geography. Geography, as its name implies, is an " earth-picturing," a presentment of earth or a portion of earth's surface in its actual form, and an indication of the influence which that form has exerted on human history or human society. To give such a picture as this of our own country, in however short and simple a fashion, is the aim of the present work. It does not pretend to furnish the learner with every possible detail of height or area. What it seeks to do is to give him some notion of what in form and surface his fatherland is like, to set before him a broad picture of these islands in the strange diversity of their structure. With this purpose it endeavours not only to point out the re-lations of the British island-group to the European continent as a whole, but to build up, step by step, the geographical skeleton of each of its parts, to trace the mountain structure of each country as defining its plains, to show how the combination of mountain and plain shapes its river system, or again how the union of all these parts furnishes a ground-work on which the national life in its local forms has necessarily shaped and moulded itself, and by whose character much of its political, and all of its industrial life has as necessarily been determined.

That such an aim has been imperfectly realised
in the present book its writers know well.   But if
their attempt to realise it does somewhat to make
the study of Geography a more living and attractive
thing the errors of detail which such a work can
hardly have escaped may well be pardoned.   For
the study of Geography, small as is the part allotted
to it in actual teaching, is one which must occupy
a foremost place in any rational system of primary
education.   When the prejudices and traditions of
our schools and schoolmasters have passed away—
as they must pass away before a truer conception
of the growth of a child's mind, and of the laws
which govern that growth—the test of right teaching
will be found in the correspondence of our instruction with the development of intellectual activity in
those whom we instruct.   The starting-point of education will be the child's first question.   And the
child's first question is about the material world in
which it finds itself.   So long as every sight and
every sound is an object of wonder, and of the
curiosity that comes of wonder, life will be a mere
string of "whats" and "whys."   With an amusing
belief in the omniscience of his elders, the child asks
why the moon changes and what are the stars, why
the river runs and where the road goes to, why the

hills are so high and what is beyond them. To answer these questions as they should be answered is to teach the little questioner Geography. Each of the divisions into which Geography breaks does its part in his training, as the picture of the earth in which he lives grows into distinctness before him. He may never hear of Physiography, but he learns in simple outline what are the forces that tell through heat and cold or wind or rain on the form of the earth, and make it the earth we see. The name of Physical Geography may never reach him, but he gets a notion of what the earth's form actually is, of the distribution of land and sea, of mountain and plain over its surface, of the relative position of continents and of countries, of the "why" rivers run and the "where" roads run to. As the structure of the world thus becomes distinct to the child he sees why races have settled, why nations lie within their boundaries, why armies have marched and battles have been fought, why commerce has taken one road or another over sea and land, and thus gathers his Historical Geography without knowing it. So as he watches how mountains divide men or rivers draw them together, how hill-line and water-parting become bounds of province and shire, how the town grows up by the stream and the port by the harbour-mouth, the boy lays the foundation of his Political Geography, though he never sees a "table of counties," or learns by rote a "list of populations."

Studied in such a fashion as this Geography would furnish a ground-work for all after instruction. It is in fact the natural starting-point for all the subjects of later training. History strikes its roots in Geography; for without a clear and vivid realization of the physical structure of a country the incidents of the life which men have lived in it can have no interest or meaning. Through History again Politics strike their roots in Geography, and many a rash generalization would have been avoided had political thinkers been trained in a knowledge of the earth they live in, and of the influence which its varying structure must needs exert on the varying political tendencies and institutions of the peoples who part its empire between them. Nor are history or politics the only studies which start naturally from such a ground-work. Physical science will claim every day a larger share in our teaching : and science finds its natural starting-point in that acquaintance with primary physics which enables a child to know how earth and the forms of earth came to be what they are. Even language, hindrance as its premature and unintelligent study has been till now to the progress of education, will form the natural consummation of instruction when it falls into its proper place as the pursuit of riper years, and is studied in its historical and geographical relations.

Such a dream of education.will doubtless long remain a dream; but even as a dream it may help us to realize the worth of Geography, and to look on the study of it in a grander as well as a more rational light than has commonly been done. It is at any rate such a dream as this that has encouraged its writers to attempt the present book. One word may be added as to their share in its authorship. Both the writers whose names appear on the title-page are responsible for the general plan of the work, as well as for the part of it which relates to England. The rest is wholly due to the second of them.

JOHN RICHARD GREEN.

CONTENTS.

ENGLAND.

CHAP.		PAGE
I. Introductory		1
II. General View of England and Wales		19
III. The Coast of England and Wales		29
IV. The Mountain Groups		40
V. The Upland Ranges		57
VI. The Plains		68
VII. The River System		75
VIII. The English Counties		1-8
IX. The Northumbrian Counties		114
X. The Counties of the Ribble and Mersey Basins		123
XI. The Counties of the Severn Basin		133

CHAP.   PAGE

XII. The Counties of the Humber Basin 145

XIII. The Counties of the Wash 162

XIV. The East Anglian Counties 172

XV. The Counties of the Thames Basin 177

XVI. The Southern Counties 194

XVII. The South-western Counties 205

WALES.

I. General View of Wales 216

II. The Welsh Counties 228

SCOTLAND.

I. Introductory 245

II. The Highlands 251

III. The Lowland Hills 269

IV. The Lowland Plain 278

V. The River System 285

CHAP.   PAGE
VI. Coast-line and Islands 302

VII. Political Divisions . .   . . : . . . 313

VIII. The Highland Counties 318

IX. The Counties of the Lowland Plain 338

X. The Counties of the Lowland Hills 351

IRELAND.

I. General View of Ireland 362

II. The Irish Counties 376

LIST OF TABLES.

PAGE

Population of the British Islands 18

ENGLAND AND WALES.

Headlands, Bays, and Islands 38

Mountains 54

Uplands 66

Rivers 106

Groups of Counties 111

Coal Fields 112

The English Counties 214

The Welsh Counties 242

SCOTLAND.

PAGE

Mountains 276

Lakes 277

Rivers 301

Bays and Headlands 311

Islands 312

The Counties 360

IRELAND.

Provinces and Counties 397

LIST OF MAPS.

PAGE

1. Map of the British Isles . . . . *Frontispiece.*

II. England and Wales. Coast-line and Water- .
    parting 21

III. The Cumbrian Hills 44

IV. The Welsh Hills 46

V. Hills of Devon and Cornwall 49

VI. England. Uplands and Plains 59

VII. England. River Systems 77

VIII. Watershed of the North Sea 81

IX. Basin of the Thames 90

X. Watershed of the Bristol Channel 96

XI. Watershed of the Irish Sea 102

XII. Counties of England and Wales 108

PAGE

XIII. Northumbrian Counties 115

XIV. Lancashire and Cheshire 124

XV. The Counties of the Severn Basin 134

XVI. The Counties of the Trent Basin 147

XVII. Yorkshire 155

XVIII. The Counties of the Wash 164

XIX. East Anglian Counties 173

XX. Counties of the Thames Basin 178

XXI. Southern Counties 195

XXII. South-Western Counties 206

XXIII. The Welsh Counties 229

XXIV. Physical Map of Scotland 244

LIST OF COLOURED MAPS.

I. Europe . . . . *to face p.* 1

II. England . . . . ,, 29

III. Scotland . . . . ,, 313

IV. Ireland . . . . ,, 362

EUROPE

By Keith Johnston, F.R.S.E.

English Miles

100   0   100   200   300   400   500

ATLANTIC

OCEAN

NORTH SEA

Great Britain

IRELAND

IRISH SEA

ENGLAND

ENGLISH CHANNEL

BAY OF BISCAY

C. Finisterre

FRANCE

SPAIN

Madrid

MEDITE...

Str. of Gibraltar

AFRICA

10   Long. W. of Greenwich   O   Long. E. of Greenwich   10

A SHORT GEOGRAPHY

OF THE

BRITISH ISLANDS.

CHAPTER I.

INTRODUCTORY.

Europe and the British Isles.—The most important group of European islands, whether in size or political consequence, is situated in the Atlantic Ocean, just off the western coast of the European mainland. This is the group of the British Isles. In number they exceed five thousand, but the bulk of them are mere masses and shelves of bare rock, and only two are of geographical importance. The first, Great Britain, is a long, narrow, irregular slip of land with jagged edges, running nearly north and south, and parted by a sea-channel from the nearest point of the mainland; the second, Ireland, lying to the westward of it, and of little more than a third its size,

is a country whose short, broad form presents a
striking contrast to that of Britain, and which resembles it in little save the torn and jagged character of
its coast-line.   The shallowness of the seas which sur-
round this island group shows that it once formed a
part of the adjacent continent.   A hundred and eighty
miles to the west of Ireland, the sea-bottom of the
Atlantic falls precipitously into a vast basin, seven
thousand feet in depth, which represents the ocean
that in some past geological epoch washed the western
shores of the European mainland.   But within this
line, to the eastward, the water suddenly shallows ; and
around the coasts of the existing islands the sea-bed is
of comparatively insignificant depth, varying from three
hundred feet deep in the English Channel to seventy
feet in the North Sea between Britain and Germany.
This slightly-sunk sea-bed, with the British Islands
which still rise above the surface of its waves, formed,
in fact, the extremity of a table-land that stretched
eastward as far as Russia, and part of which still exists
in the plain of Northern Europe.   It was a slight depression of this table-land that allowed the sea to
cover a great part of its surface, and by forming the
inlets now known as the English, the Irish, and the
Northern Channels, to break what remained into
islands, and to part these islands from the rest of
the European Continent.

   Their Physical Resemblance.—It is this iden-
tity with the Continent which explains the physical
structure of the British Isles.   If we look at any general
map of Europe, we see that the northern part of the

European continent consists of a vast plain, stretching from the Steppes of Asia to the British Channel;
while at right angles to this plain, in a direction from
north to south, runs out a peninsula, that of Sleswig
and Jutland, whose line is carried northward by the
Danish isles and the mountain mass which constitutes Scandinavia.   Both these features of Northern
Europe are repeated in the British Islands.   The great
northern plain is prolonged across the bed of the
North Sea to form the southern half of Great Britain
and the centre of Ireland.   On the other hand, the
characteristics of the Scandinavian peninsula repro-
duce themselves in the northern projection of Britain,
in its long, narrow, and rugged form, and in the nature
of its mountain-ranges, which strike down the western
side of the island from north to south, fronting the
ocean on their western side for the most part in a
steep wall of heights, and sinking slowly to the eastern
sea in more gradual slopes, down which are thrown
all the more important rivers of the country.

PHYSICAL CHARACTERISTICS OF THE BRITISH
ISLANDS.—But though the two great islands of the
British group are thus linked together by their relation
to the Continent, their severance from it has been the
cause of many points of *contrast*.

I. Climate.—Their insular position has a marked
effect on their climate.   It gives them warmth, it
preserves them from extremes of heat and cold, and
it produces an abundance of moisture.

(*a*.) The British Isles are much *warmer* tha
other countries, save Norway, whether in Europe or

America, which lie so far to the northward. The mean temperature of Ireland in the fifty-second degree of latitude is the same as that of the United States of America in the thirty-second degree of latitude, though at this point the latter are situated more than two thousand miles nearer the equator.

(*b.*) Again, in these islands great *changes of temperature* are unknown. Scotland lies at about the same distance from the equator as Central Russia; but while the mean difference between summer heat and winter cold is in Scotland only nineteen degrees, in Central Russia it amounts to forty-eight degrees. This mild and equable climate is caused by the fact that all winds coming to the British Isles must first pass over the sea, and in the case of the most frequent of the British winds, the west wind, over great spaces of sea. They are therefore cooler in summer and warmer in winter than winds which travel over tracts of land parched and burnt by the sun or frozen by bitter cold. As all parts of the British Isles are bathed in the same sea-winds, so all share alike in the advantages they bring; and there is scarcely any difference between the winter climate of the southern coast of Britain and that of the Shetlands in the far north.

(*c.*) But as the winds cross the water they gather moisture, which falls in *rain* when they touch the colder land. Hence the rainfall of these islands is greater than that of the rest of Europe, its average being over three feet. The prevailing wind is from the west, the number of days on which a westerly

wind blows exceeding the number of days on which
the wind is easterly in a proportion of forty-five to
twenty-eight. This west wind, laden with the moist
vapours of the ocean which it has traversed, first
breaks upon Ireland ; and hence the climate of this
island is damper than that of the rest of the British
group. Ireland has an average of two hundred and
eight rainy days in the year. As its western mountain-
chains, however, are comparatively low, the rain-
clouds are not effectually caught by them, and so dis-
tribute their moisture more evenly over the whole
surface of the country than in the Island of Britain,
the difference between the rainfall on the eastern and
western coasts of Ireland being only nine inches. In
Britain, on the other hand, the mountains of the west
are high enough to intercept a large proportion of the
moisture borne along by this wind, and in some districts of this part of the country the rainfall reaches
a ratio of more than seven feet a year. A belt of
rainy country thus extends along Western Britain,
but this excess of waters, through the mountainous
nature of the ground, is carried seaward in rapid
torrents, which are of little avail for navigation or
the promotion of fertility. In the eastern plains of
Britain the rainfall still remains considerable, amounting to from one and a half to three feet in the year ;
but the undulating surface of the country, while it
hinders its accumulation in marshes, conduces to the
gathering of its waters together into rivers of greater
size and more gradual descent than on the west, and
thus renders them of greater utility to the soil.

(*d.*) This moderate climate and abundant supply of moisture co-operate to promote the *fertility* of the British Islands. Wheat can be cultivated even in their northernmost regions; and over a large part of Britain it may be grown at a height of 1000 feet above the sea-level; while in the mountains of the north, where this is impossible, it is replaced by hardier grains, such as oats and barley. But over the bulk of Ireland and in large districts of Britain, especially in the western half of it, where the rainfall is injurious to crops, the ground is more profitably used for pastures, which are nowhere richer than here.

II. **Physical Structure.**—Important as are these results of their climate, the British Isles owe yet more to the peculiarities of their physical structure. In hardly any part of the world is such a variety of geological formations to be seen as in Britain; almost every kind of rock, from the oldest to the most recent deposits, is in turn brought to the surface in some part of these islands. The result of this is that the mineral treasures which exist in some of these formations are here easily accessible, and that in spite of the mineral·resources of America, Africa, and Australia, Britain still remains the most productive mining country in the world. Its rocks contain iron, tin, copper, lead, zinc, and salt; but they contain a yet more valuable possession in their vast beds of coal. The coal-measures of Britain, which extend over an area of nearly 3000 square miles, have been the centre from which the industrial movement of modern times in Europe and America has received its main impulse. In Britain itself they are the source of vast industries; for manufactures of iron, steel, cutlery, hardware, glass, pottery, cotton, wool, and other textile fabrics cluster round the coal-beds wherever they are found. From Ireland, indeed, coal is absent; and it is through the want of it that, in spite of some mineral resources, this country remains without manufactures, save that of linen, which flourishes in those northern parts which adjoin the coal-beds of North Britain. But this contrast only brings into stronger light the physical characteristics which have given its mineral wealth to Britain itself.

III. **Position.**—As the mining and manufacturing industry of Britain is a result of its peculiar physical structure, so the commerce of these islands is in great part a result of their peculiar geographical position. Placed as they are at the north-western angle of Europe, they command the whole commerce of its northern countries, as it passes through the North Sea and the British Channel; while they lie in the most direct route for vessels bound from America to Europe. London in fact is placed at what is very nearly the geometrical centre of those masses of land which make up the earth-surface of the globe; and is thus, more than any city of the world, the natural point of convergence for its different lines of navigation. The position too of these islands in the ocean gives them over neighbouring European countries the advantage of tides high enough to carry large vessels up the estuaries and harbours, whose abundance along their coasts furnishes another element of commercial importance.

IV. **Political Geography.**—Again, the peculiarities of the physical structure of the British Islands have told on their political history.

1. *Political and Social Differences of Ireland and Great Britain.*—To them are due the social and political differences which so long parted Ireland from Great Britain, and which still retard their practical union. Ireland is distinguished from Great Britain by its size, as well as by the characteristics both of its position and of its physical structure. Its area amounts to 30,000 square miles ; its greatest length is 290 miles, and its greatest breadth 175 miles. It is thus little more than a third as large as the Island of Great Britain. It differs from it as strongly in position as in structure. While it is separated from America by the whole width of the Atlantic Ocean, it is cut off from Europe by the greater island to the westward of it. This isolated position has to a great extent protected it from foreign invasion, and has preserved in the country its ancient inhabitants with but little change. But at the same time the island has been in great measure shut out from direct contact with the general civilizing influences of Europe, and it was only in comparatively late times that its wandering tribes were brought within the full scope of European civilization. The remarkable unity of its physical structure, so strongly in contrast with the variety of that of Great Britain, has had even greater social results. It is in fact reflected in the unvaried charac-

ter of its industry. The centre of the island forms a broad, level plain, broken only by lakes, and traversed by one large river, round which runs a circle of hills and low mountains, which form a wide belt along the northern and southern shores, and a narrower belt along the eastern and western coasts. The moisture of the Irish climate, which results from the position of this island in the Atlantic Ocean, produces a constant rainfall which makes pasture more profitable than tillage; and this vast central plain has in all ages been mainly a grazing-ground for cattle. Manufactures, save in the north, there are none. The uniform character of the rocks, from which coal is absent, and which contain metals in but small quantities, have prevented any general growth of manufacturing industry, and thus restricted the great bulk of the inhabitants to agricultural employments.

2. *Political Division of Great Britain itself.*— While physical characteristics thus distinguish Ireland from Great Britain as a whole, they have also been the chief causes which have brought about the division of Great Britain itself into three separate realms, and which have to a great extent determined the social and political character of each of these portions. Taken as a whole, Britain covers an area of 84,000 square miles, its length is about 600 miles from north to south, while its breadth varies from 33 to 367 miles. But it is only in comparatively recent times that the island has become a single nation. For many centuries it was divided

into three separate countries, that of Scotland to the north, and Wales to the west, while the southern mass of the island bore the name of England.

3. *Political Geography of Scotland.*—Scotland owed its separate existence partly to the form of Great Britain, and partly to the physical differences between the northern and the bulk of the southern parts of the island. The disproportion between the length and the breadth of Britain threw a great obstacle in the way of political unity in times when communications were slow and difficult. Still greater obstacles were interposed by the physical difference between the districts to the north of the Cheviots— the hills which formed the frontier between Scotland and England—and those to the south of them. While the latter consisted for the most part of low hills and open plains, Scotland was little more than a continuous mass of high mountains separated by a strip of level ground from another mass of bare and lofty hills which extend to the border. This natural division of the country into Highlands, or the mountain district, and Lowlands, or the district of the hills and the plains, was reflected in the twofold character of its population, the Gael of the Highlands being distinguished from the Englishman of the Lowlands by differences of race and speech which long held them apart as two separate peoples. Within each district too the broken character of the country tended to promote the division of its inhabitants into separate clans or bodies obeying their special chieftains, and hindered all efforts to bring them to any

real national oneness. The barrenness also of their soil, and the inclemency of their climate long doomed the Scotch to extreme poverty; while their position, which cut them off from the civilizing influence of Europe, kept them in a state of barbarism. On the other hand, these hardships helped in creating a thrift and endurance which enabled the smaller country to hold its own against its greater neighbour. Scotland is only about half the size of England, its area being but 24,000 square miles, its extreme length 286, and its breadth varying from 33 to 160 miles; while in wealth and population it was greatly inferior. But in spite of this disproportion it preserved its independence till a peaceful union was brought about by one of the Scotch kings mounting the English throne. Since this union its physical structure has again played a great part in its social history. Whilst the general progress of civilization has removed much of the dissociating effects of its mountainous character, the discovery of rich coal-beds in the level strip between Highlands and Lowlands has given birth to manufactures and large towns, and, with the upgrowth of agriculture in the south through the energy of its people, has turned the poverty of the country into comparative wealth.

4. *Political Geography of Wales.*—The severance of Wales from the mass of the country was due to only one of the circumstances which promoted the separate existence of Scotland. Wales consists of a tract of rugged, mountainous country projecting into the Irish Sea from the western coast of Southern

Britain, whose boundary-line on the eastern side runs
where these mountains sink into the plains of which
Central England is made up. It thus resembles Scotland in its general features, save that its mountains
nowhere rise so high as those of Scotland, and that
it is far smaller, being only a seventh the size of
England, and covering an area of but 7,400 square
miles. It was in fact a small mountain fastness in
which the older, or Celtic, peoples of Northern
Britain took refuge when the more open country was
conquered by the English invaders, and where they
long preserved their independence. Wales, like
Scotland, remained for ages poor and barbarous ; it
was in the same way cut off by England from contact
with civilized Europe, while the bleak moorlands and
slaty hills which cover most of its surface long
afforded only a thin pasture for small sheep and
cattle, and supported but a few inhabitants. But
again, like Scotland, Wales has in more recent times
owed a vast social change to its physical structure.
In the northern and southern parts of the country rich
beds of coal, iron, and slate have been found and
worked ; and in these districts the general solitude
and poverty of the country is now exchanged for busy
industry and a teeming population.

5. *Political Geography of England.*—The rest of
the island, south of the Cheviots and west of Wales,
formed the kingdom of England. It is far the
largest of the three, for it is 350 miles long, and has
an area of 50,000 square miles. In appearance it at
once resembles and differs from the neighbouring

countries, for its mountains are less bold than those
of Scotland or Wales, while its plains are less monotonous than those of Ireland.   In fertility and general
wealth it is far more favoured than any.   Its surface
is more varied; it consists mainly of wide undulating
plains, but these are broken by low uplands in the
south and east, and by masses of mountains and
hills in the north and south-west.   The undulating
character of most of its surface not only facilitates
communication and road-making, but promotes the
gathering together of the waters of the country into
streams and rivers.   No other part of the British
Isles can compare with England in the complete-
ness of its system of rivers, which are spread over
the face of the land in an order more perfect than
that of almost any other European country.   Unlike
the mere mountain torrents which are so common in
Scotland and Wales, these rivers form a network of
navigable waters and carry fertility to all its plains.
The fruitfulness of England is aided by its climate, a
climate less damp than that of Ireland and less cold
than that of Scotland, but sufficiently varied in the
amount of its rainfall to allow the growth of grain
over the eastern half of the country, while it provides
rich pastures through the western.   To these advantages we may add that of mineral wealth, in a vast
extent of coal-beds and rich deposits of iron and
other metals, as well as salt-mines and beds of valuable clay.   But much as it owes to its physical
structure, England owes hardly less to its geographical
position.   It lies closer to Europe than either Scotland,

Ireland, or Wales, and thus possesses greater advantages for trade and commerce, while its harbours are situated in more sheltered seas than those of its fellow-countries.

Both its structure and position have combined to shape its political history. Easy communication with the Continent has brought England within reach of its civilizing influences, as it has left her open to descents of invaders, who have driven before them her older inhabitants to the mountains of the north and west. Thus it has come about that the Englishmen of Southern Britain are more recent incomers into the land than the Gael of the Highlands or the Cymry of Wales. That their conquest of the country was possible was due in great measure to the absence of mountain-barriers such as checked their progress in these latter districts; but it is the absence of such barriers which in later days enabled government to act easily over the whole face of the land, and soon drew its various tribes together into one people and a highly organized realm. With such advantages of structure and position, and with the far greater wealth and population that came of them, it was inevitable that England should in the long run gather the neighbouring realms round it, and that it should form by far the most important element in the political body which has resulted from their union—the United Kingdom of Great Britain and Ireland.

General Results of the Character of the British Group.—From these special consequences of the position and structure of each of its parts on their character and history, we may now proceed to sum up some of the more general consequences of their structure and position as a whole. The fact that they have been cut off from the Continent of Europe by a belt of sea which forms them into islands, has produced most important results on the character of these countries, on the growth of their civilization, on the training of their peoples, and on the part they have played in the world's history.

(1.) The strip of sea which runs between the British Isles and the Continent has greatly influenced *the progress of their inhabitants from barbarism to civilization.* Though foreign conquerors have in early ages been able to cross over the sea to Britain; the English Channel and the North Sea have long formed a barrier broad enough to prevent perpetual invasion and disquiet from without, and to give such a measure of security as was needed for the well-being of the British peoples. But on the other hand these seas have not been wide enough to shut out the civilizing influences of Europe; foreign trade, and wealth, and knowledge have been within easy reach of its people. And in some cases the channel which parts these islands from Europe has even invited civilization to their shores. It has made them a safe place of shelter for men driven out from their own lands by poverty, by misgovernment, or by persecution, and wave after wave of immigrants have thus been brought to our coasts, who have carried with them a knowledge of arts and manufactures.

(2.) Again, it is owing to the sea that Great Britain has won its place as the *greatest naval power* of the world. This greatness has been the natural consequence of men living in an island where all trade and communication with other peoples is wholly impossible except that which is carried on across seas both dangerous and stormy, and where sailors are trained to all kinds of peril and hardship. The training of the people to this seafaring life is made more complete and widespread by the very form of the islands, long, narrow, with shores deeply indented by creeks and bays, so that no spot in the very heart of Great Britain could be more than 120 miles from the sea, and most of its inhabitants must at all times have been familiar with the ocean.

(3.) The agricultural, manufacturing, and commercial industries which the British Isles owe to their physical structure and European position enable them to maintain a very large *population* for the size of the country, more than thirty-one millions of inhabitants. The people are scattered indeed over the country in a most unequal way—sometimes, in a great city like London, over 9000 are crowded together in a square mile, while again, in some desert mountainous district of Northern Britain, scarcely thirteen persons can be found in the same space. But this distribution of population is mainly determined by the distribution of mineral wealth, especially of coal with its accompanying manufactures : in all mining districts the people are thickly gathered together in large towns,

while agricultural districts with their rural villages
and hamlets are but thinly peopled. The various
countries in fact support a population exactly in
proportion to the amount of their coal-measures.
England has an average of 422 people to the square
mile, Wales of 164, Scotland of 112. In Ireland there
are 161 persons to the square mile, an unusually large
number for a mainly agricultural country.

(4.) The most important result of this large population
and of the narrow space of land within which it is
confined is our *colonization* of distant countries, and the
enormous consequent growth of the political importance of Great Britain as an imperial power. As the
inhabitants of these islands have grown too many for
the narrow limits of their home, they have gone
out from it to America, Africa, Asia, and Australia, to
conquer new lands, or to found settlements of people
speaking the English tongue, and taking with them
English laws and customs. These colonies have so
increased in number and extent, that while the whole
area of the British Isles (121,692 square miles) does
not represent more than one-sixteen-hundredth part
of the surface of the globe, and is not equal to one-
thirtieth part of the continent of Europe, the dominion
of Britain now extends over more than one-seventh of
the dry land of the world, that is, over a territory
equal to sixty-seven times the area of the British Isles,
and containing more than 200 millions of inhabitants.
The English language is spoken by 88 millions of
people as their mother-tongue, and if we include
men of other races who understand the language, by

C

at least 100 millions—a number which promises soon
to increase two- or three-fold, as commercial interests
extend yet further.   It is owing then in great part to
geographical circumstances that the political influence
of the English race has become one of the most
important facts of the world's history.

Plan of the Book.—In our survey of the
British islands, we shall follow the order which is
determined by the social and material importance of
their various parts.   First we shall describe the geographical and political divisions of England with its
dependency of Wales, then those of Scotland, and
lastly those of Ireland.

Country.	Area in square miles.	Population.
England	51,005	21,495.131
Wales . . . . :	7398	1,217,135
Scotland	30,463	3.360,018
Great Britain . . .	88,866	26,072,284
Ireland	32,523	5.412,377
Isle of Man and Channel Islands	303	144,638
United Kingdom . .	121,692	31,629,299

CHAPTER II.

Character of the Country.—From our survey of the general structure of the British Islands we now therefore turn to survey that portion of them which we know as Southern Britain, or the countries of England and Wales. The first object on which our view must be fixed is the character of the land itself, for it is this that determines the inner history, progress, and temper of a people by the daily influence it exerts on all their enterprises. We must learn therefore to know the actual surface of Southern Britain—where it is broken by rocky mountains, where its rivers have formed valleys rich with corn and fruits, and where its broad uplands furnish pastures for flocks. Nor is any of this knowledge useless. We get to understand the true importance of these facts as we watch how from age to age the inexorable lie of the land has compelled very different races and generations of men to live and build their cities by the same rivers in sheltered valleys or in fertile plains; to till the ground in the

C 2

same fruitful districts; to make their roads along the same lines marked out by nature, sometimes across level plains, sometimes threading narrow valleys between the hills, now turning aside to reach some river-ford, and then rising over a mountain pass where the line of hills dips so as to form a gap in the forbidding heights. We learn why it is that the great English roads are as old as the earliest times told of by any history of England, and why so many towns and even villages have a story that goes back to the first beginnings of our people. And many things hard to understand in history become easy as we know the outer circumstances which led to the events of which we read.

England and Wales, taken together, form a very irregular triangle, lying between three seas, the North Sea, the Irish Sea, and the English Channel, and having but one land boundary, that on the north, where England adjoins Scotland. The border line is here marked by the short chain of the *Cheviot Mountains,* and by those rivers which flow from either extremity of this line of hills to the sea, the *Solway Firth* on the south-east, and the *Tweed* on the north-east (see Frontispiece). At first sight the tract within this triangle may seem a mere confusion of hills, and plains, and rivers, in which it is impossible to distinguish any sort of plan. Nevertheless, it is certain that order can be found in all this seeming disorder if we begin by seizing clearly the great general divisions marked by nature herself.

The Water-parting.—For this purpose, the first

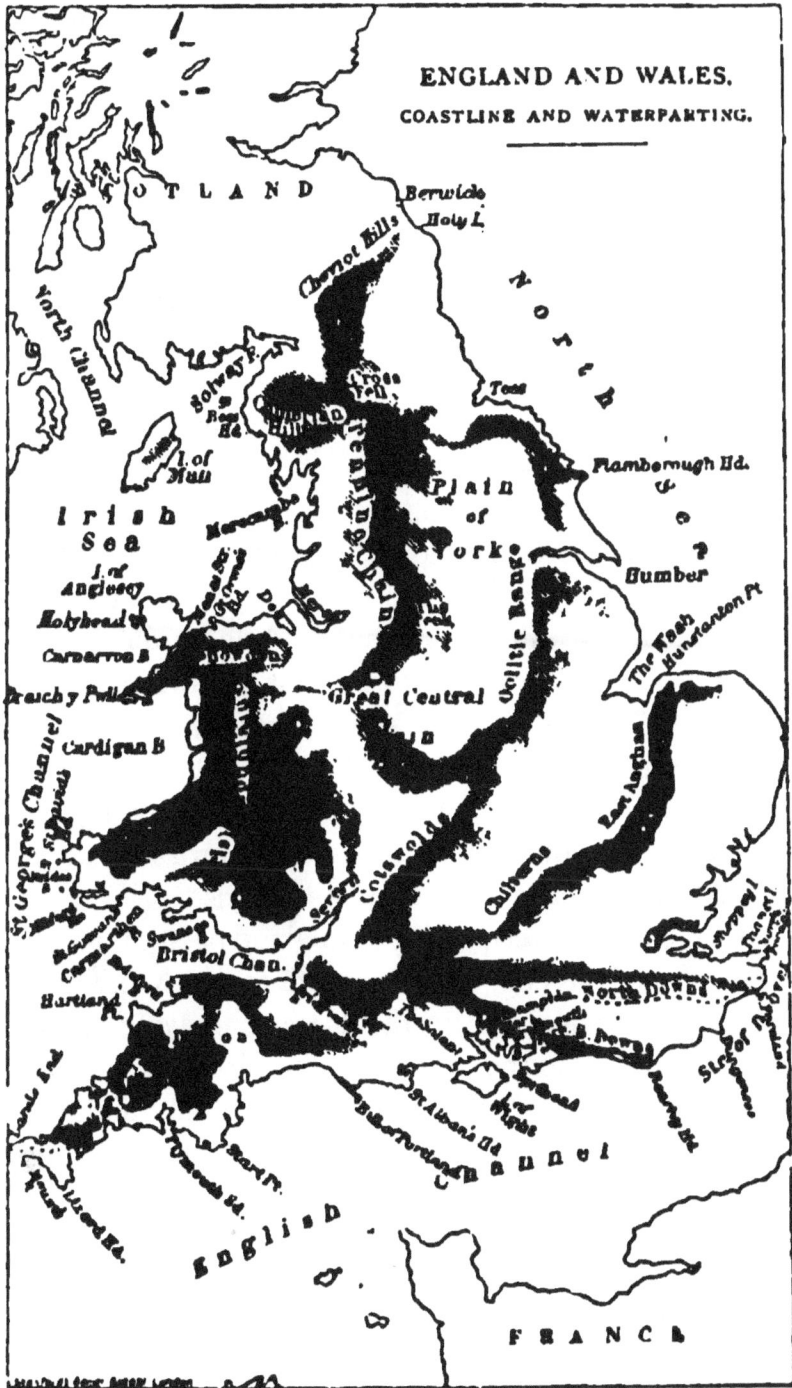

ENGLAND AND WALES.

COASTLINE AND WATERPARTING.

line of demarcation to lay hold of is the *Water-part-ing* of England. The water-parting of any country is that line of rising ground which parts its rivers one from another. As the springs of the rivers lie on different slopes of the hills or undulating ground where they take their rise, they are thrown downwards to the sea in different directions, while the higher ground between them forms the water-parting. The character of the country traversed by the line which indicates it may vary very much. Sometimes this line is drawn along the tops of high hills; sometimes it crosses a great upland where the slope of the ground can hardly be discerned, or a low broken country full of little hills and gentle undulations. But difficult as it may be to trace it, it always exists. Among all these differing kinds of scenery are to be found springs of rivers, and between the little streams which flow from them lies a line of slightly higher ground which compels them to travel down the slope on that side where they first took their rise.

Now the chief water-parting of England passes through the centre of the country from north to south, dividing it into two unequal parts, the larger part lying to the east, the smaller to the west. The line drawn along this water-parting is carried through three totally different kinds of country. For nearly half its length it passes down the centre of a chain of high moors and mountains, called the *Pennine Range*, which runs due south for 200 miles from the Cheviot Hills to the Peak of Derbyshire; a long unbroken line of bleak and desolate table-lands, covered with heather and thin mountain grass, whose solitude is only broken by the sheep grazing on the broad slopes, and by the sound of stone-quarrying. From the Peak it runs south through the broad and fertile *plain of central England*, a plain broken here and there by low hills. Finally its southernmost course lies across an upland country of swelling *downs*, such as the Cotswolds, covered in great part with open pasture lands. This line therefore passes directly through all the great features of English scenery, its moors and mountains, its great plains, and its uplands.

Contrast between Eastern and Western England.—But mountains, plains, and uplands do not lie equally distributed over the surface of the country; on the contrary, the water-parting marks a very distinct division between them (see p. 21). To the west of it are gathered huge mountain-masses and wild moorlands, formed of very ancient, hard, and rugged rocks. To the east, on the other hand, lie broad plains of alluvial soil, and fertile river valleys, parted from one another by uplands of chalk and other soft rocks, which have been worn into gently rounded outlines by wind and weather so as to form low swelling downs, bearing no resemblance to the mountains of the west.

View of Western England.—If we look from the water-parting in detail to west and east, we shall see distinctly the difference between the character and scenery of these two divisions of the country. Let us suppose a man to be standing on some high hill at the southern extremity of the Pennine chain in

Derbyshire, and looking westward over the mountains
which lie between him and the Irish Sea.   He will
see three distinct *mountain groups*, like three mighty
buttresses, thrown out from the western coast into
the sea, and parted from one another by deep and
wide bays.

First, to the west and south-west, lies the great
central group of the *Welsh mountains*—a vast tract of
moorland broken by hill-ranges rising one behind
another with bare and rugged outlines, till they reach
their greatest height by the sea-coast in the peaks of
Snowdon, the loftiest mountain of southern Britain.
The whole of this Welsh group forms a peninsula,
very short in comparison to its great breadth, lying
between the Irish Sea and the Bristol Channel, and
throwing out westward two spurs of mountainous rock.
In size it is immensely larger than the other groups ;
its mountain ranges are more bold and rugged, its
peaks higher, its valleys wilder, its bleak moorlands
more extensive, its soil more barren.

Looking now to the north of this group, the spectator will see a second mass of mountains parted from
the first by a broad plain which has swept round
from eastern England and stretches along the shore of
the Irish Sea.   Beyond this plain to northward lies a
small circular mass of high hills, set between two deep
indentations of the sea to north and south.   This
little group of the *Cumbrian Hills* rising up out of
the sea, with its lines of mountains shutting in narrow
wooded valleys, each one of which contains its own lake,
forms the most beautiful spot of all English scenery.

Far away in the extreme south-west lies a third group of highlands, formed by the heights of *Devon and Cornwall.* Here a long narrow arm of rock stretches out seawards between the Bristol and the English Channels, and presents a treeless tract of high moorlands, very bleak and barren, forming sometimes a mere waste of boggy soil, sometimes covered by a thin mountain pasture. The moors are broken by high hills of granite, heavy and rounded in form, thrown up from the table-lands ; and a wall of bold cliffs lines the coast.

View of Eastern England.—To gain such a view as this from any one point is of course impossible. But the fancy of it may help us to realise the true character of western England, and how it differs from the east. For let us now suppose the same spectator to turn his back on the mass of mountains that stretches away behind him to the Irish Sea, and to look out over England to the east and south-east with such an imaginary telescope as will carry his eye as far as the North Sea, to which the whole country dips by slight degrees. Just before him stretch interminable low undulations, forming a vast plain narrowing to the south, a plain traversed by great rivers, and bounded by a flat-topped escarpment facing to the north-west and rising boldly above the level ground. This escarpment is the edge of a table-land, known as the *Oolitic uplands,* which slowly sinks on its eastern side to a lesser plain. From the further side of this second plain rises another steep escarpment, that of the *Chilterns* and *East Anglian Heights,*

which forms the edge of a second table-land, that
in its turn sinks gently downwards on its eastern
side, and has its slopes hollowed out by little river-
valleys.   These two escarpments, one after another,
run right across the scene from the shores of the
English Channel in the south to the brink of the
North Sea in the north-east and east.   Beyond them
to the south-east lies the great basin of the
river Thames, and yet further off in the extreme
distance, the observer would catch sight of two
narrow straight lines of hills, the *North* and *South
Downs*, thrown out towards the Straits of Dover,
and forming boundary walls between yet smaller
plains which open out on the shores of the south-
eastern sea.

These upland ranges are, therefore, four in number; and taken together their form is roughly that
of a vast misshapen hand spread out over the
face of eastern and southern England, and inclosing
between its fingers broad tracts of low-lying country
plentifully watered by rivers, and covered with fields
of corn and wheat, fruit orchards, and wooded pasture-
land.   The whole surface of the land on this eastern
side falls by a long and gradual descent to the ocean,
while on the west the very reverse happens, and it
rises to its greatest height in the mountains which
front the western sea.

Results of this Contrast.—This disposition of
mountains and plains into two distinct divisions of
southern Britain has been the cause of several
important contrasts between the two parts of the

country. Some of these we have already mentioned, but must briefly notice once more.

(*a*.) The wild western moors and hills are naturally unfitted for tillage, while they are often rich in coal, copper, lead, tin, or valuable stone. Hence the chief, though not all the mining and manufacturing industries of England are carried on in its northern and western districts, while the broad plains and river valleys of the south and east are employed in agriculture and pasture.

(*b*.) There is also a difference in climate between the two parts. The mountains of the west act as a screen to protect the eastern plains on that side where they are in danger of excessive rainfall which would injure the crops. As heavy masses of clouds drive up from the Atlantic Ocean, they are caught by the mountain heights, and robbed of the greater part of their moisture before they pass on to the plains. Thus while the barren hills are constantly clothed in mist, the fields beyond them receive only the needful amount of water.

(*c*.) There are also historical and political differences between the two divisions of Southern Britain, for their peoples are in great part of different race and habits. The low fruitful grounds of the east lay near to Europe, a tempting prey open to enemies from abroad. As wave after wave of foreign invasion swept over the land, the older inhabitants were driven back for refuge to the mountain fastnesses, where they could long defend themselves from conquest. And here the descendants of these tribes still live, so that the mountainous districts of the west are now peopled by representatives of the oldest races of the land, while in the lowlands of the east the descendants of the later conquering peoples are to be found. We have already seen how the physical differences between the eastern and western parts of the island have told on their population and wealth.

Such a view of England as we have taken from its water-parting has enabled us to realize its general character, and the relation of its various parts. We may now therefore proceed to a more detailed survey of the country, and examine in turn its coast, its mountains, its uplands, its plains, and its system of rivers.

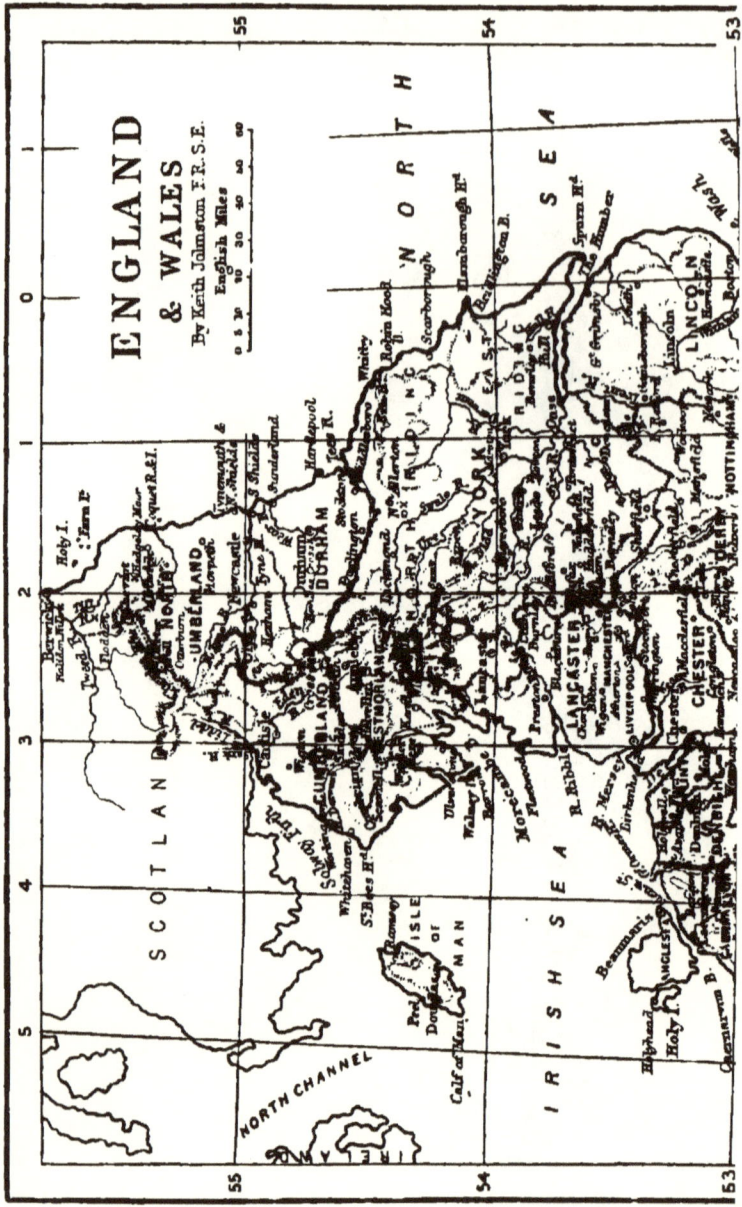

ENGLAND & WALES

By Keith Johnston F.R.S.E.

English Miles

NORTH SEA

SCOTLAND

IRISH SEA

IRELAND

NORTH CHANNEL

ISLE OF MAN

Dominion of MAN

Calf of Man

NORTHUMBERLAND

DURHAM

CUMBERLAND

WESTMORLAND

LANCASTER

YORK

NORTH RIDING

EAST RIDING

LINCOLN

CHESTER

NOTTINGHAM

LIVERPOOL

ANGLESEA

Holy I.

Holyhead

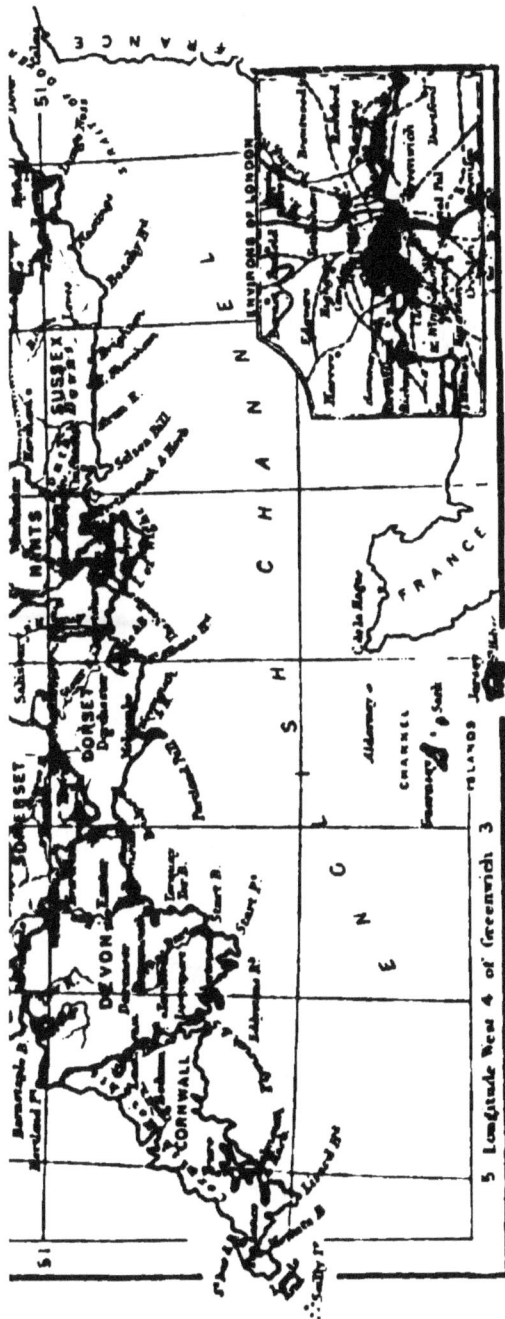

CHAPTER III.

Importance of Coast-line.—In any country having a sea-board, but especially in an island, the character of the coast-line is of the highest importance. As the inner structure of the country determines to a great extent its inner history and character, so its shores, its harbours, and its river estuaries influence its outer relations, its intercourse with the world, its trade and commercial development. It is possible for a maritime country to have a very great length of shore unbroken by rivers or by indentations of the sea ; and a land thus destitute of harbours, and shut out from intercourse with foreign nations remains poor and barbarous. Other countries with a far smaller extent of shore have their coasts so deeply cut by inlets of the sea or by large rivers that from the earliest times trade, and consequently civilization, have been drawn to them. Those parts of the world which are now most thickly peopled and prosperous are those whose coasts are most abundantly furnished with harbours.

The Character of the Coast of any country

depends on the inner structure of the land itself. Our general view over Southern Britain will enable us to understand how the shores of England and Wales are the necessary outcome of the mode in which the island is built up, and of the resistance which its rocks are able to offer to the seas which dash against them on every side. The irregular triangle composed of these two countries opposes three different sides to the sea (see p. 21). The eastern side, from Berwick to the South Foreland, measures in a straight line 350 miles; the southern, from the South Foreland to Land's End, 325 miles; the western, from Land's End to Berwick, 425 miles. But these shores are so deeply hollowed out by bays and inlets of the sea, and so cut and indented by estuaries and harbours, that the total length of the coast-line amounts to 1,800 miles. There is no country of the same size in the world, save Greece, which is marked by an outline so varied and irregular as that of Britain.

This irregular character, however, is not equally true of every part of its shores, for it varies in the different coasts according to the character of the land itself on that side. Let us first shortly sum up the marked peculiarities of each coast, and see how these correspond with the inner structure of the country.

The *western coast* consists mainly of four deep and wide bays, parted by projecting buttresses of land thrust out into the sea. Bold cliffs and masses of mountainous rocks bound the greater part of the shore, rising sharply above the sea with stern and rugged outlines. These cliffs are formed of very

old and hard rocks, which are strong enough to resist
the effort of waves and weather to wear them down to
more even and gentler forms, and have thus preserved
their rugged features in spite of the sea which beats
against them, a deep and strong sea with very high
tides and great waves.

The *eastern coast*, on the contrary, is a long, low,
monotonous line of level shore, sloping to the south-
east, little broken by headlands, cliffs, bays, or inlets
of the sea, its only openings being formed by the
estuaries of the rivers which here empty themselves
into the sea. Where there are cliffs, the rocks are
generally lower than those on the west side, being
softer and easily worn away by weather. The shore
consists chiefly of gravel, clay, and sand, with long
reaches of desolate marshy ground, parted from one
another by lines of chalk cliffs worn by the waves
into gentle curves. From these low banks the bottom
of the sea slopes away very gradually, so that the bed
of the ocean for some distance is shallow, full of
shoals and sand-banks, and the tides are low and
regular in their quiet ebb and flow.

The *southern coast* of England combines the peculiarities of both the eastern and the western coasts.
That half of it which lies to the eastward is an uneven
line of low shore—slightly broken—first a bit of chalk
cliff, then a stretch of level clay ; a second line of low
cliff, and a second tract of flat clay. But that half
which lies to the west begins to form itself into bays.
Two wide curves appear, such as those which lie on
the western side of the island, and in these two bays

the character of the cliffs changes, till the bold outline
of the mountainous west is clearly marked.   Further,
in passing from east to west along the English
Channel which washes the southern coast, we find
that the sea grows deeper and stronger as we get
farther west, till we reach the great waves of the
Atlantic.

The *cause of this difference between the coasts* is made
clear if we remember the difference which exists
between the eastern and western divisions of the
country, and how the whole land lies in an inclined
plane, sloping downwards from west to east.   It is
the rugged mountain masses of the west which form
those massive peninsulas that distinguish the western
coast, while the eastern shore is simply made up of
the terminations of the low plains and the upland
ranges which alternate with them on this side.   The
gradual fall of the land from west to east also explains
why it is that the openings cut in the east coast are
made by river estuaries.   The streams which rise
in the higher land to the west are thrown down the
hill-sides to the low ground of the east, and as
they pass through the soft sands and marshes of
the shore to empty themselves into the North Sea,
they hollow out those wide channels which break the
regularity of the coast-line.

But after this general survey we must study the
coasts in greater detail.

I. The **east coast** lies on the North Sea, sloping
slightly from north-west to south-east, from Berwick
upon Tweed to the South Foreland.

(*a.*) There is but one *island* of any note along this shore—*Holy Island*, near the north-eastern coast. The other isles are merely tracts of low swamp lying off the marshy coast of the Thames estuary, such as *Foulness* and *Sheppey* Islands.

(*b.*) The coast is very simple in structure ; it is divided into nearly equal parts by four *river openings*, which increase in size as we proceed southwards. The northern half is but slightly broken by two smaller estuaries, the mouths of the *Tees* and the *Humber*, which form harbours opening out towards Norway, Denmark, and Germany. The southern half is more irregular, owing to two larger estuaries, the *Wash* and the mouth of the *Thames*. The Wash is formed by the washing in of the sea on a low marsh where the openings of several rivers meet together, and is absolutely useless for shipping. The whole trade of the south is therefore gathered into the greatest of all the estuaries, the mouth of the Thames, which by its size as well as by its position so close to Europe is the most important harbour in England.

(*c.*) The *headlands* of the east coast are few in number. They are formed at three distinct points, where the upland ranges of the east abut upon the sea in cliffs composed mainly of chalk rocks which break the level shore. The cliffs which terminate the oolitic uplands extend from the mouth of the Tees as far as *Flamborough Head*, above the marshes of the Humber. *Hunstanton Point* marks the extremity of the East Anglian Heights which line the shore south of the Wash ; and between the Thames estuary and the straits of Dover are the *North* and *South Forelands*, and the *Dover* cliffs, formed by the North Downs.

II. The **south coast** of England lies, from the South Foreland to the Land's End, on the English Channel,—a belt of sea which widens westward into the Atlantic Ocean, and on the east opens through the Straits of Dover into the North Sea. These narrow straits, only twenty miles wide at one place, part England from the nearest Continental country, France.

(*a.*) There is but one large *island* adjoining the shore, the *Isle of Wight*, situated near the central part of the coast, with but a narrow passage between it and the mainland, called on the north-east *Spithead*, and on the north-west the *Solent.*

At some distance from the Land's End, the most south-western point of the British Islands, is the small cluster of the *Scilly Islands.*

(*b.*) The coast to westward and eastward of its central point is, as we have seen, of very distinct character. From the cliffs of Dover to the Isle of Wight the slightly curved shore is only marked by the low clay spit of *Dunge Ness,* and the chalk cliffs of *Beachy Head,* at the extremity of the upland range of the South Downs. On the other hand, to the west of the Isle of Wight, the bays formed by the curvature of the shore are parted by several projecting *headlands* —*S. Alban's Head,* the *Bill of Portland,* a long spur of rock striking out seawards, and *Start Point,* a promontory of very old and hard rock, in strong contrast to the soft chalk of Beachy Head. The extreme south-western part of the coast is formed

by two abrupt masses of volcanic rock thrust out
into the sea from the Cornish heights, *Lizard Head*,
and the *Land's End*, which enclose between them
Mount's Bay.

(*c.*) This coast contains three very great and
important *harbours.* Two of these, *Portsmouth Harbour* and *Southampton Water*, lie directly behind the
Isle of Wight. Portsmouth, completely landlocked,
forms a magnificent port, and is the great naval
arsenal of Britain : Southampton has a large trade
with Europe and the colonies. The third harbour,
Plymouth Sound, in the S.W., is the chief naval arsenal
of the country.

III. The **west coast** of England, extending from
the Land's End to the Solway Firth, is parted from
Ireland by a narrow belt of sea. In its widest central
part this is known as the Irish Sea, from whence it
opens out into the Atlantic Ocean by the narrower
passage of St. George's Channel on the south, and
the strait of the North Channel on the north.

(*a.*) Two *islands* lie in this sea. The *Isle of Man*,
half-way between England and Ireland, is thrown
across the centre of the Irish Sea. The *Isle of
Anglesea* lies close to the coast of Wales so as to
seem part of the mainland ; the *Menai Strait*, which
cuts it off from Wales, being so narrow as to be crossed
by a suspension-bridge.

(*b.*) The four *bays* of the western coast differ much
in character. While two of them, those to the north
and the south, are narrow, and terminate in river
estuaries, the two central ones consist of wide inlets

of the sea formed by the curvature of the shore. All alike, however, lie between rocky and mountainous peninsulas running out into the sea in the same south-westerly direction. The northernmost opening in the coast is the estuary of the *Solway Firth*, which severs England from Scotland, and is bounded only on the southern side by English land.

To the south of this firth lies a deep curve formed by the Irish Sea as it washes in upon a belt of low sandy shore between the mountains of Cumberland on the north, and the mountains of Wales on the south. The low coast-line which curves inwards between these heights is interrupted by *Morecambe Bay*, lying under the Cumbrian hills, and is further cleft by many wide river estuaries, of which that of the *Mersey* is the most important. These rivers form ports of great value, whose westerly position gives them the chief trade with Ireland and America; the harbour at the estuary of the Mersey is only second in importance to that of the Thames itself.

The next great bend of the shore forms *Cardigan Bay*, and is of a very different character from the last. It lies encircled by the Welsh hills, the line of its curve being sharply broken to the north by a narrow ridge of high rocks thrust out from the mountain masses which bound it on this side; while to the east the hills sweep in a semicircle round the bay, sinking lower as they near its southern boundaries. Owing to these encircling hills and to the evenness of the coast-line, which is only broken by very small rivers, this great bay is useless for trade and shipping. To north

and south of it lie lesser inlets of the sea.   On the
north is *Carnarvon Bay*, between the Isle of Anglesea
and the rocky tongue of land which forms the peninsula of Carnarvon: on the south *St. Bride's Bay*
lies at the very extremity of the peninsula.   Beside
St. Bride's Bay a long creek running inland from the
sea forms the only harbour of this district, *Milford
Haven*.

The fourth opening of the west coast differs very
much from the last two in form and character.
Just opposite to the estuary of the Thames, a deep arm
of the sea, the *Bristol Channel*, is driven between the
mountainous country of Wales on the north, and the
highlands of the long peninsula which lies to the south
with its bold and rocky shores.   Constantly narrowing
as it stretches more inland, this channel finally passes
into a great river estuary, the mouth of the *Severn*,
which here breaks the coast-line.   The harbour of
Bristol, formed in this estuary, has a large shipping
trade, chiefly across St. George's Channel with Ireland.
The shores of the Bristol Channel are curved in its
wider part into lesser bays, *Swansea* and *Carmarthen
Bays* on the north, *Bideford* and *Bridgewater Bays*
on the south.

(*c.*) The *headlands* of the west coast are all formed
by projecting spurs of rock thrown out from the three
mountain groups which front the sea on this side.
The chief among them are *St. Bee's Head*, a height
jutting out from the Cumbrian Hills between the
Solway Firth and the great bay formed by the Irish
Sea; *Great Orme's Head*, on the Welsh coast to the

south of this bay; *Holyhead* on Holyhead Island, and *Braich-y-Pwll*, on either side of Carnarvon Bay; *St. David's Head*, at the southern extremity of Cardigan Bay; *St. Gowan's Head* and *Hartland Point*, guarding the opening of the Bristol Channel, one on the northern, the other on the southern side. The extreme south-western point of English soil, the *Land's End*, is formed by an abrupt mass of granite rock in which the long projecting peninsula of the south terminates.

We must now turn from this survey of the coast, and of the political and commercial results of its conformation, to a survey of the land itself, and of its inner structure. The most prominent features of this structure are its mountain-groups, and it is to a study of these that we may proceed in the next chapter.

HEADLANDS.

East Coast.		South Coast.		West Coast.	
	Ft.		Ft.		Ft.
Flamborough Head	450	Dunge Ness	92	Hartland Point	350
Hunstanton Point	60	Beachy Head	564	S. Gowan's Head	166
North Foreland	184	S. Alban's Head	440	S. David's Head	100
South Foreland	370	Portland Bill	30	Braich-y-Pwll	584
Dover Cliffs	469	Start Point	204	Holyhead	719
		The Lizard	224	Great Orme's Head	678
		Land's End	60	S. Bee's Head	456

ISLANDS.

East Coast.	South Coast.	West Coast.
Holy Island.	Isle of Wight.	Isle of Anglesea.
The Farne Islands.	Scilly Islands.	Isle of Man.
Sheppey Island.		

BAYS AND ESTUARIES.

East Coast.	South Coast.	West Coast.
The Tees. The Humber. The Wash. The Thames.	Portsmouth Harbour. Southampton Water. Plymouth Sound. Mount's Bay.	Bideford Bay. Bristol Channel. Swansea Bay. Caermarthen Bay. Milford Haven. S. Bride's Bay. Cardigan Bay. Estuary of the Dee. Estuary of the Mersey. Morecambe Bay. Solway Firth.

CHAPTER IV.

Importance of Mountains.—As we looked out from the water-parting over western England, we saw the country broken up into three great masses of mountains. We must now examine these three districts more in detail.

The mountains of any country are among its most important features. By them the *river-system* of the land is in great measure determined; they influence the *climate* for good or evil to the tillage of the soil; by the character and mode of upheaval of their rocks it is decided whether *mines* of precious metals are to be brought to the surface or buried unknown in the earth; and on their position in the country, and the degree of their steepness and wildness, depends the question whether *inland trade and commerce* can be easy and profitable, or whether it will be checked by overpowering difficulties of communication. In England the mountain groups are so disposed as to give many advantages to the country. They send

down to the plains abundant rivers, they shield the
island from excessive rain, they contain great stores of
coal and mineral treasure, while they allow a free
communication between nearly every part of the
island, so that wealth and civilization may be easily
diffused.

The mountains of England lie in four distinct groups
—(1) the *Pennine Range*, (2) the *Cumbrian Hills*, (3)
the *Welsh Mountains*, and (4) the *Highlands of Devon
and Cornwall.*

I. The **Pennine Range** (see p. 59) is a vast succession of moorlands and masses of hills stretching
southward from the Cheviot Hills in a long unbroken
line to the heart of England, and forming a kind of
backbone to the country. Throughout its whole course
it lies nearer to the Irish Sea than to the North Sea,
falling rapidly down to the western coast, but sinking
on the east to the broad plain of York. Its general
height is between 1,000 and 2,000 feet—but it rises to
nearly 3,000 feet in *Cross Fell* in the north, where the
mountain-ridge is wild and narrow. Further south
it broadens, and from its great table-lands throws up
a group of mountains, *Mickle Fell, Whernside, Ingleborough*, and many others on the eastern side, while on
the west offshoots are sent out towards the Irish Sea, so
that the moors here reach forty miles in breadth.
Again narrowing a little to the *Peak*, a region of moorlands and rounded hills, the range dies down gradually
to the central plain of England, after having reached
a length of two hundred miles. A line of mountain
heights so wild and steep and of so great a length

forms almost a wall of separation between the east
and west of northern England, shutting them out to
a great extent from direct communication with each
other, and driving all traffic between England and
Scotland along two routes which alone remain open to
it, the lowlands of the coast on the west, and the plain
of York on the east. The scenery of the moors remains throughout wild and bleak; heather and thin
grass alone cover the uplands, and the rough hills are
in many places pierced by deep caverns formed in
their rocks of mountain limestone, into which the hill-
torrents plunge to travel for a time under ground.
In the straitened river valleys alone can shelter be
found for some little agriculture, and a few small
villages. All besides is silent and deserted.

There is one point, however, towards the southern part of the range where the character of the
moorland slopes is wholly changed. Here the upheaval of the lower rocks has rent asunder a vast
coalfield which once stretched across this part of
England, and rich beds of coal and iron lie extended
on the slopes of the range to east and west, where
they have become the centre of the greatest mining
and manufacturing industries of England. The
working of iron and steel is carried on alike on both
sides of the range, but besides these there are special
industries which belong respectively to the eastern and
western slopes. The eastern or *Yorkshire* moors form
the great seat of the *woollen trade*, where the whole
of the English manufacture of cloth and worsted is
centred. On the western or *Lancashire* moors is the

cotton manufacture, whence England draws its entire supply for internal and foreign trade. These industries have gathered round them, in what were once moorland solitudes, the densest population in all England, and have crowded the hill-sides with towns half hidden under the cloud of smoke sent up from the tall chimneys of innumerable factories, blast-furnaces, and coal-pits.

II. The **Cumbrian Group** lies on the western side of the Pennine Chain, to which it is bound by a spur of high moorlands thrown across not far from Whernside. This connecting belt of moors dips midway between the two mountain masses to form the pass of *Shap Fell*, over which a road climbs by which all trade passes between the western parts of England and Scotland. To the north of Shap Fell the Cumbrian Mountains are severed from the Pennines by the broad valley of the river *Eden;* to the south of it by the narrow glen formed by the river *Lune.* The two bays into which these rivers open, *Solway Firth* and *Morecambe Bay*, enclose the Cumbrian group to north and south.

Within these boundaries rises a small, compact, circular mass of mountains, very steep on their northern and western faces, but sinking gradually to Morecambe Bay on the south in long and gentle slopes. The centre of the group is marked by one of its greatest mountains, *Helvellyn*, 3,000 feet high; the northern borders by *Skiddaw;* the greatest western heights by *Scafell*, the loftiest mountain in England, 3,162 feet high, and by *Bowfell* lying near it. To the south is *Coniston Old Man,* whence the hills gradually fall to the sea in the peninsula of *Furness.*

The great feature which distinguishes this district from the rest of England is the number of its *lakes,* which are the more remarkable as being the only large

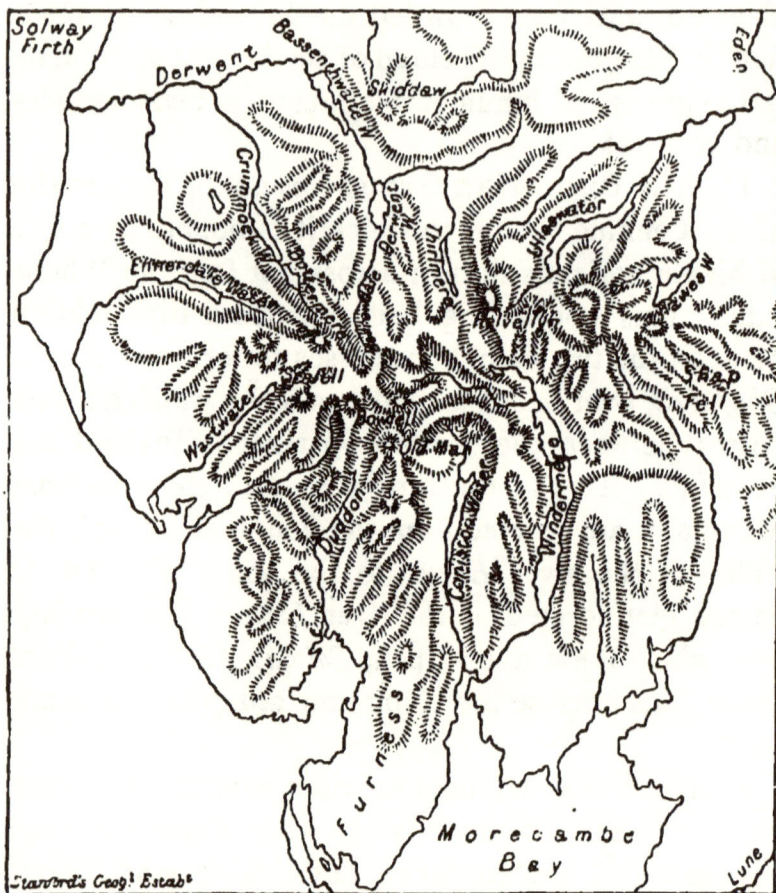

CUMBRIAN HILLS.

sheets of water throughout the whole country, and so give to this region the name of the *Lake District.* These lakes lie in a somewhat regular order, occupying

long narrow valleys, which are ranged round the centre
of the group, Helvellyn, and radiate from it outwards
in all directions like the spokes of a wheel. The largest,
Windermere, points south; *Hawes Water* and *Ulleswater*, north-east; *Thirlmere*, north; *Derwentwater*
and *Bassenthwaite Water*, north-west; *Buttermere* and
Crummock Water, also north-west; *Ennerdale Water*,
west; *Wastwater*, south-west; and *Coniston Water*
nearly south. In the midst of them, a little to the west
of Helvellyn, lies the high central valley of *Borrowdale*,
shut in on every side by closely encircling hills, and
very famous for the beauty of its scenery. The steep
sides of the mountains as they descend to the sea
are cut by many other valleys, deep, narrow, and
wild, worn by the mountain torrents flowing from the
upper lakes. But the only river of any size is the
Derwent, which, by means of its tributaries, drains
all the lakes of the north-west, itself taking its rise in
Derwentwater, close to Borrowdale.

The population of this district is scanty : its industries are limited to sheep-farming, and the working of
one or two lead mines, and of a coal-bed on the sea-
shore. But the character of the rocks, soil, and climate,
which have limited its means of wealth, have at the
same time given to it a beauty in which it surpasses all
other parts of England.

III. The mountains of the **Welsh Group** form
a broad peninsula bounded on the north by the Irish
Sea, and on the south by the Bristol Channel. This
peninsula extends westward to St. George's Channel,
where it sends out in a south-westerly direction

two spurs of mountainous rock, of which the upper
one is narrow and abrupt, the lower one broader and
longer, and which include between them Cardigan
Bay. The low island of Anglesea, lying close to the
land, has the look of a third projection pointing to the north-west. The narrow north-eastern part

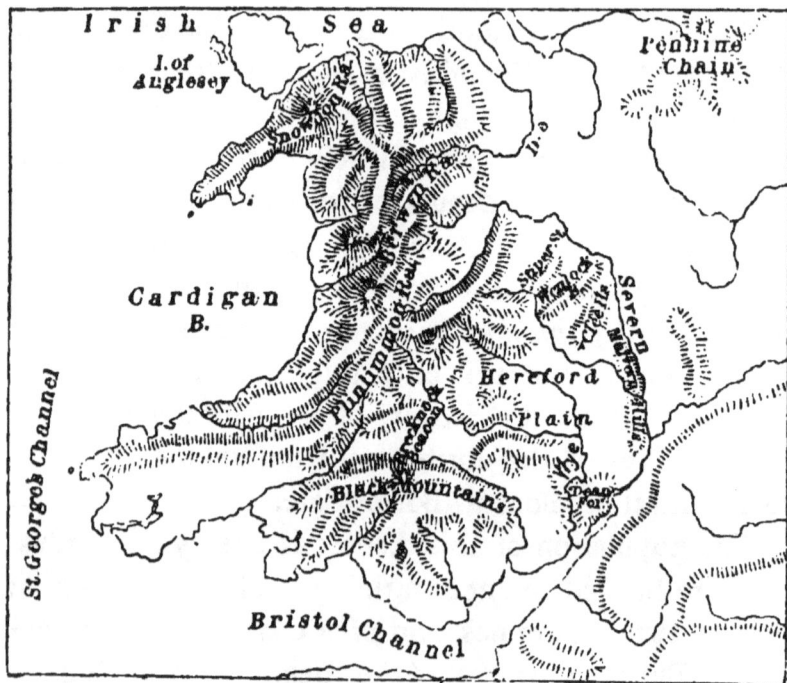

WELSH HILLS.

of this mountain group is bounded on the east by the
river *Dee* and the wide plain stretching beyond it to
the Pennine chain ; the broader south-eastern portion
is cut off from the plain of Central England by the
narrow valley of the river *Severn*. These two valleys
form a continuation to the southward of that long

belt of low ground which extends from the head of the Solway Firth along the western side of the Pennine Chain, save where it is broken by the pass of Shap-Fell, and which marks the eastern boundary of the ancient rocks known as Cambrian and Silurian, the first-formed land of Southern Britain (see p. 21).

The whole of the mountainous district which lies to the west of these rivers formerly belonged to the principality of *Wales;* and though part of it has now been taken into England, the greater part still retains the older name. For the sake of clearness, the whole group must first be considered under the general name of the Welsh group; afterwards it will be possible to divide between the part now included in England, and that which is still known as Wales.

The structure of these mountains can be best understood by observing the direction of their chief ranges. Of these there are four, besides a tract of very broken country which lies to the south-west. Three of the mountain ranges run nearly in the same direction, from north-east to south-west—(1) the *Snowdon Range;* (2) the *Berwyn Mountains;* (3) the *Plinlimmon Range;* (4) the fourth and most southerly range, the *Black Mountains* and *Brecknock Beacon*, lie due east and west; (5) the broken country made up of *Hereford Plain* with its surrounding circle of irregular heights forms the easternmost part of the mountain group.

The mountain districts of Wales are all alike in character, consisting as they do of extensive tracts of wild moorland broken by masses of hills, which rise in the north-west in Snowdon to 3570 feet, a greater height than any other Welsh summit, and form a scenery of extreme grandeur and boldness. The moors are studded with small tarns instead of lakes; and give rise to a number of rivers that dash downwards to the plains in steep waterfalls and rapid torrents, the chief of these being the *Severn* and the *Wye*, which take their rise near the centre of the group. The whole mountainous district is less remarkable, however, for the height of its hills than for the variety of its wild and picturesque scenery, the great beauty of its glens and mountain gorges, and the abundance of its tarns and running waters. But its rough moorlands and slaty hills are barren and thinly-peopled, save by the borders of the Bristol Channel, where the southern slopes of the Brecknock Beacons and Black Mountains contain a bed of coal and iron that stretches down to the sea-coast, and here, as on the Pennine Moors, a sudden change takes place, and crowded population and ceaseless activity take the place of silence and desolation.

The fifth division, the broken plain of *Hereford*, which has been taken from Wales to add to England, is the only naturally fertile district lying to the west of the Severn. It is broken by gently-sloping hills, and surrounded on all sides by mountainous country. On the south lie the Black Mountains, with their outlying ridges; on the west the high moorland country which stretches along the eastern slopes of Plinlimmon; on the north long lines of hills radiating northwards,—*Stiper Stones, Long Mynd, Wenlock Edge* and the *Clee Hills;* and on the east a ridge of steep heights which border the valley of the Severn, the chief of these being the *Malvern Hills* and *Dean Forest.* The wild and rugged character of the Welsh mountains is only preserved in the western boundaries of this district, the eastern hills being much lower and more gentle, while the broad plain, which forms in fact the fruitful basin of the river *Wye,* is rich with orchards and highly-cultivated ground.

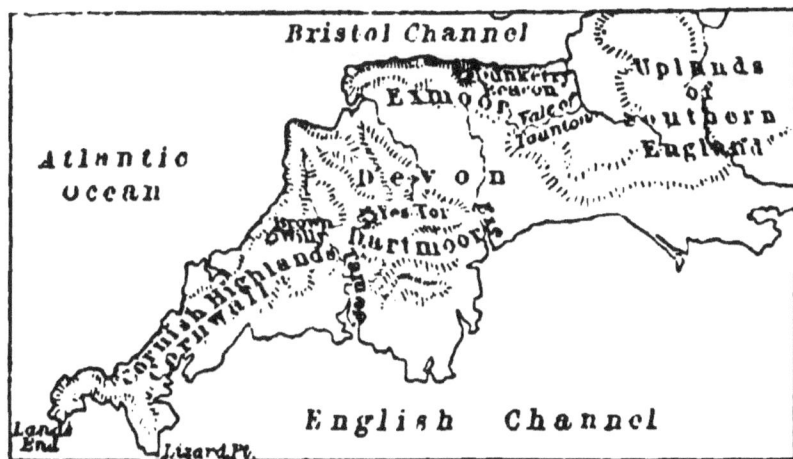

HILLS OF DEVON AND CORNWALL.

IV. The peninsula of **Devon and Cornwall**, which reaches out to the Atlantic between the Bristol and the English Channels, is cut off from the uplands of southern England by the valley of the river *Parret* and by the *Vale of Taunton*—valleys which in fact form a continuation to the south of the great depression of the Severn valley, and like it divide the ancient

E

granites, schists, and slates of the western mountains from the newer rocks of the east. (See p. 21.)

The peninsula which lies to westward of these valleys differs in character from the mountain groups to northward of it, but it is closely allied to the peninsula of Brittany in France, which forms the opposite shore of the Channel that has severed these two districts óne from another. On both sides of the sea there is the same geological structure, the same rugged and indented coast, the same climate, the same race of inhabitants; the very headlands in which the peninsulas terminate bear the same name of Land's End, or Fin-des-terres.

The *Devonian Heights* rise on the southern shores of the British Channel in the lofty tableland of *Exmoor* and the *Quantock Hills* to the east of it. Exmoor itself forms a barren and treeless moorland where gloomy stretches of bog alternate with wide reaches of sheep-pasture, cleft by wooded ravines. The hills with which its surface is broken rise from 1,200 feet to a height of 1,700 feet in *Dunkerry Beacon.* On its northern side the tableland falls to the sea in steep cliffs, while on the southern side the ground sinks gradually towards the wide bay which lies between Start Point and Portland Bill, to which the rivers of the moorlands, such as the *Exe*, are thrown down.

To westward of the broad plain which forms the lower valley of the Exe, and inclosed between that river and the *Tamar*, rises *Dartmoor*, a second tract of moorland between 1,000 and 2,000 feet high, more extensive than the first, and yet more rugged and barren in scenery, without grass or trees, and like Exmoor, covered by wide stretches of bog. It rises to its greatest height of 2,040 feet on its western side in the summit of *Yes Tor*, whence it slopes by degrees to the English Channel, forming a projecting mass thrust out against the waves, which terminates in Start Point, and is bounded to east and west by deep harbours, the mouth of the Exe, and Plymouth Sound. This mass of heights is the centre where a number of small rivers take their rise, and flow in all directions, such as the *Plym*, the *Dart*, the *Teign*, and the *Tawe* —all streams of little importance. It is a richer district than Exmoor, for its rocks of slate and granite contain mines of tin, copper, lead, and iron; and quarries of granite, marble, and limestone, besides beds of porcelain clay.

To westward of the Tamar lie the *Cornish High-lands*, a succession of granite moorlands broken by hills much lower than those of Dartmoor and Exmoor. These tablelands and hills, beginning on the shores of Bideford Bay at Hartland Point, narrow so as to form a single line of heights, with a rapid slope on either side to the sea, and run to the south and south-west as far as the high granite headland of the *Land's End* and the volcanic rocks of *Lizard Point*, while the bare cliffs of the Scilly Islands beyond show how far the peninsula once stretched. The chief moun-tains of Cornwall are *Brown Willy*, 1,368 feet high, and close to it *Rough Tor* rising out of a tableland of dreary, treeless waste, covered with bog.

Beneath the gloomy and barren surface of Cornwall

valuable minerals lie buried. Its rocks contain some of the very few tin-mines in the world, which, as far as history takes us back, have given it a most important foreign trade. Copper is also more abundant in its rocks than anywhere else in Great Britain, and besides these it possesses lead-mines and great granite quarries. Here alone in the peninsula do we find the country tolerably well peopled, for a large mining population has been gathered to its south-western end, where the mineral wealth is most abundant.

Comparison of these Mountain Groups.— The four mountain groups of the west coast have certain points of *likeness*. (*a.*) They are alike composed of those ancient Cambrian, Silurian, and Devonian rocks whose summits rose as islets in mid-ocean while the rocks of eastern England yet lay hidden in its depths. (*b.*) They share in the same poverty of soil, in the same excessive moisture of climate, and consequently in the same scarcity of tillage and scantiness of population. (*c.*) Their sources of wealth and industry are the same— sheep-grazing, stone quarrying, and the working of mines of coal and metal.

But they have also their points of *unlikeness*, each one differing from all the rest in (*a*) relative position, (*b*) in form and size, (*c*) in scenery, and (*d*) in the quantity of their mineral wealth (see p. 21.)

The Pennine chain lies apart from the rest by its inland position ; by its form of a long narrow wedge driven down the centre of the country, it divides

between two parts· of England so as to check communication, while on the other hand the mountains of
the coast form a barrier, not between two portions of the
land, but between land and sea. But as the Pennines
traverse the narrowest and most remote part of England,
and run in a direction from north to south, the barrier
which they form does not wholly cut off any part of the
island from all other parts, and their influence in retard-
ing the spread of free intercourse is slight. In this
they differ from the Cheviot Mountains, which run
across the island from east to west, and so break it into
distinct parts, which remained separate kingdoms for
many hundred years. The Pennine Moors differ also
from the other mountain groups in the greater extent
of their coal-fields, and in the variety of the manufactures peculiar to them, for example, wool, cotton,
silk, pottery, hardware, and cutlery.

The three groups which rest on the sea contrast
strongly with one another in the four points mentioned.
They differ (a) in relative position. The Cumbrian
group lies west of the high mass of the Pennine Chain ;
the hills of Devon and Cornwall west of the low chalk
downs of southern England ; and the Welsh Mountains
west of the broken central plain of England. (b.) Again,
their forms are different ; the Cumbrian group is a small,
round, compact ring of heights, projecting very little
into the sea ; the Devonian group, a long, irregular,
and narrow peninsula thrust out into the ocean for
a great distance ; and the Welsh group, a much larger
mass than the other two, very broad, reaching out
into the sea farther than the northern group but not so

far as the southern, and sending out from its western shores two projecting tongues of land. (*c.*) The character of the scenery varies in each group. The Cumbrian Mountains are steep and high, lying close together, and parted only by narrow valleys, with a lake lying in the centre of each valley. Devon and Cornwall, on the contrary, form a mountain plain raised high above the sea level, chiefly composed of a succession of bleak moorlands, broken here and there by hills, but having no fertile valleys, no lakes, and no steep mountain heights. The Welsh group combines in itself the character of the other two. In the north and south it has mountains like those of the Cumbrian group, but far steeper, higher, bolder, and more extensive ; while in the centre and south-east there are high moors and mountain plains like those of Devon or Cornwall, but far wilder and more desolate. Owing to its greater size, however, it differs in a very marked way from the other groups in the extent and importance of its river-system. (*d.*) There is a difference, too, in the mineral wealth of the three districts. The Cumbrian Hills are poor in minerals, possessing only a bed of coal and a few lead-mines ; Cornwall has its rare metals of tin and copper ; while Wales contains, besides lead and copper, one of the greatest coal and iron fields in Britain.

MOUNTAINS.

The Pennine Range.

Cross Fell	2,892 feet.
Mickle Fell	2,591 ,,

Whernside . . . . 2,414 feet.
Ingleborough . . . . . . 2,373 „
Pen-y-gent . . . . 2,273 „
Pendle Hill . . . . 1,831 „
The Peak, Kinderscaut . . . 1,981 „
The Peak, Lord's Scat . . . 1,816 „
Axe Edge Hill . . . . . . 1,810 „
Weaver Hill . . . . 1,205 .,

The Cumbrian Group.

Scafell . . . . 3,162 feet.
Helvellyn . . . . 3,118 „
Skiddaw . . . . 3,054 „
Fairfield . . . . 2,863 „
Great Gavel . . . . . . 2,949 „
Bow Fell . . . . 2,960 „
Pillar . . . . 2,932 „
Saddle Back . . . . . . 2,847 „
Coniston Old Man . . . . 2,577 „

The Welsh Group.

Snowdon . . . . 3,570 feet.
Carnedd Llewellyn . . . . 3,482 „
Aran Mowddwy . . . . . 2,972 „
Cader Idris . . . . 2,929 „
Moel Shiabod . . . . . . 2,865 „
Brecknock Beacon . . . . 2,910 „
Cradle (Brecknock) . . . . 2,630 „
Plinlimmon . . . . 2,469 „
Radnor Forest . . . . . . 2,166 .,
Stiper Stones (highest point) . 1,759 „
Long Mynd . . . . 1,674 „

Caradoc Hills . . . . . . 1,200 feet.
Wenlock Edge . . . .   "
Clee Hills . . . . 1,805 "

The Devonian Group.

Exmoor, Dunkerry Beacon . . 1,707 feet

Dartmoor, Yes Tor . . . . . 2,040 "
Cornish Highlands, Brown Willy 1,368 ,,
      "   "   Rough Tor 1,296 ,,

CHAPTER V.

Scenery of Eastern England.—If we now look from the water-parting over *eastern* England, a very different scene lies before us from that presented by the mountainous districts of the west. Instead of rugged mountains and moorlands covered with heath, bog, or scanty grass, we see lines of low hills, undulating downs, and broad river plains with their corn-fields, grass-lands, and orchards of fruit. Instead of wide reaches of deserted moors, broken here and there by the busy life of the coal-fields with their crowded manufacturing cities, we find a country with the population spread evenly over its surface in innumerable villages and quiet country towns lying by the banks of rivers. The activity, wealth, and unsightliness of the mining districts are exchanged for the stillness of agricultural life. This contrast is brought about by the different character of the ground. The soft rocks of' the east are without the mineral wealth, as they are without the torn and distorted character of the western mountains. The greatest elevations which we see are

simply undulations of the ground, formed chiefly by
long lines of uplands dipping and rising again a they
traverse the great plains.   In very few parts of eastern
England do the hills rise higher than from 500 to 900
feet, that is, to half the height of the chief mountains
of Devon and Cornwall, or about a quarter the height
of the chief Cumbrian Mountains; and their gentle
slopes contrast strongly with the bold outlines of the
volcanic rocks and masses of hard slate which are
found in the west.

Uplands.—Before examining the uplands of eastern
England we must be very careful to avoid all confusion
between upland and mountain.   An *upland* is a tract
of country often very extensive, raised sometimes many
hundred feet, sometimes only 200 feet, above the
level of the sea.   Its surface may be quite smooth,
or varied by undulating ground—it may be extremely
fertile, or comparatively bleak and barren, according to the character of the soil.   Again, it is
possible for the upland to be low and even, and
the plain beside it very much broken by hills and
slightly raised, thus making the distinction between
them to be well-nigh lost in parts; while in other
places the plain may be perfectly flat with the
high ground rising sharply from it like a wall so
as to form a steep *escarpment.*   Thus the uplands
which cut through the broken plains of central and
eastern England are sufficient to add variety and beauty
to the scenery, without giving those strongly-marked
features to the country which distinguish the west.
The soft clays and chalk which compose them are

ENGLAND.

UPLANDS AND PLAINS.

Flamborough H^d

PLAIN OF YORK

PLAIN OF

CHESHIRE

CENTRAL PLAIN

THE FENS

Hunstanton Pt

THAMES VALLEY

The Weald

New Forest

Purbeck I^d

ENGLISH CHANNEL

Beachy H^d

PENNINE CHAIN.	OOLITIC RANGE.	CHALK RANGE.
1. Cross Fell.	1. Lyme Regis.	
2. Whernside.	2. Cotswolds.	1. Dorset Heights.
3. Pen-y-gent.	3. Edge Hills.	2. Salisbury Plain.
	4. Northampton Uplands.	3. Marlborough Downs.
CENTRAL PLAIN.	5. Rockingham Forest.	4. Hampshire Downs.
1. Cannock Chase.	6. Lincoln Uplands.	5. White Horse Hill.
2. Dudley Hills.	7. Lincoln Wolds.	6. Ilsley Down.
	8. York Wolds.	7. Chiltern Hills.
	9. York Moors.	8. East Anglian Uplands.

worn away by time and weather into gently rounded lines of hills, and no part of them is wholly unfit for either agriculture or pasture.

The Upland Ranges.—Though the features of upland and plain are sometimes so blended together in eastern England as to make the line of separation between them very slight, yet the ranges which form the skeleton of the country on this side of the water-parting can be distinctly traced. As the west of England has its four mountain masses, so the east has its four ranges of uplands, as the framework on which it is built up. Of these we may call the first the **Oolitic Range** ; while the three others, that is to say, the *Chilterns* and *East Anglian range*, the *North Downs*, and the *South Downs*, compose the **ranges of the Chalk.** These ranges, however, are disposed in an order wholly different from that of the west. Beginning on the shores of the English Channel, where they are closely linked together, they stretch out to the north-east and east, making the form of a great misshapen hand—a hand having the palm of a dwarf, and the fingers of a giant. The northernmost range is the longest, and contains the steepest heights ; the other ranges decrease gradually in length and in height to the South Downs, which is the least of all in size, so that the whole surface of the land forms a gradual slope from the oolitic uplands to the sea-shore. The oolitic range and the first of the chalk ranges are alike in their general direction ; both of them traverse England from sea to sea, and both form on their western sides a tolerably steep escarpment, while their eastern slopes descend very gently and gradually to the plains below. The two more southern branches of the chalk uplands, the North and South Downs, differ from them in being much narrower and shorter lines of hilly country, more uniform on both their faces.

I. **The Oolitic Range** of uplands takes its name from the character of the limestones of which it is in great part composed, whose stones are made up of a multitude of little round egg-like particles, *oon* being the Greek word for an egg. These uplands form a long chain of heights traversing England from the cliffs which line the shores of the English Channel on either side of *Lyme Regis* in the south, to the mouth of the Tees on the north-eastern coast. The whole range taken by itself is in form something like a huge sickle, having its handle resting on the English Channel, and the flattened end of its blade on the German Ocean. Its great length is made up of a number of different parts. The southern extremity, which forms the handle of the sickle, begins in the rocks by Lyme Regis. Thence an irregular line of low broken heights extends northwards towards the estuary of the Severn, where they suddenly change in character as they emerge into the *Cotswold Hills*. These hills form a range of high tablelands with a very steep escarpment looking out over the Severn valley, and a gradual slope on their eastern side ; their soil will grow little but a thin grass, so that their whole industry consists of sheep-grazing. The long line of the Cotswolds as it trends round to the north-east forms the beginning of the blade of the sickle, and sinking lower and lower, soon passes into the smaller and more broken *Edge Hills*. From the Edge Hills the *Northampton Heights* stretch in a north-easterly direction, and present a-low broad tableland, scarcely raised above the plain to west of it. Their north-eastern portion is known by the name of *Rockingham Forest*, from the extremity of which a narrow belt of the same gently undulating country, called the *Lincoln Uplands*, runs northwards to the Humber, where it unites with a second very short line of uplands formed of chalk, the *Lincoln Wolds*.

To the north of the Humber both these ranges are prolonged together in a neck of high land called the *York Wolds*, composed for the most part of chalk. They are for some distance narrowed to right and left by the *Plain of York* and the *Holderness* marsh; but suddenly the heights expand on either side—eastward to the sea-shore at *Flamborough Head*, the beginning of a long wall of sea-cliffs reaching to the Tees, and westward in the semi-circular line of the *Hambledon Hills*. The broad northern end of the whole range is formed by the *York Moors*, a tract of wild moorland which runs east and west, and unites the Hambledon Hills with the sea-cliffs of the eastern coast. These moors are the highest and wildest part of the whole range, and therefore of the whole of eastern England. They are bleak and solitary, with their steep and barren heights, their scattered sheep-farms, their lonely tracts of gorse, and their clusters of stunted trees lying in sheltered hollows. Their southern side is

deeply cut by narrow river valleys which run down to the
low ground of the *Vale of Pickering*, shut in between
the Moors and the Wolds: while their northern face
towards the Tees forms the steep line of the *Cleveland
Hills*, famous for producing the best iron in England,
and for being at present the only mining and manufacturing district of the English uplands.

Throughout the whole extent of the range, made up
as it is of all these parts, the western face is steeper
than the eastern, and forms a great escarpment which
overlooks the plains of central England and of
York. This escarpment, like the uplands which it
bounds, rises to its greatest height in its two opposite
extremities, the York Moors, and the Cotswold Hills,
while the lower central part of the range dips to the
plain, and like it is covered with corn and wheat and
richer pasture-lands. But the western escarpment of
the uplands is throughout its whole length clearly
defined by the river valleys which run along its base.
The valley of the *Ouse* bounds the Hambledon Hills
and York Wolds; the valley of the *Trent* lies below
the Lincoln heights; the *Avon* passes under the Northampton uplands and Edge Hills to the Cotswolds,
below which lies the *Severn*.

The oolitic range is important as marking a very
distinct natural division between two parts of England. It forms the eastern boundary of the manufacturing districts, which nowhere cross the limits
set by it. To the west and north-west of the
range, therefore, lie the whole industrial wealth and
activity of England, its centres of dense population,

and its commercial cities, inhabited chiefly by in-
telligent and trained artisans, classes of rising importance in English political and social life. To
the south-east of the range, on the contrary, are the
agricultural districts of England, containing but one
great centre of crowded population in the valley of
the Thames.

II. **The Chalk Ranges.**—If we now look at England from this oolitic range eastward we shall see that
what remains of it is an irregular triangle bounded
by the English Channel, the North Sea, and a line
drawn from the Wash to the Channel about Lyme.
All this vast triangle is occupied by the *chalk uplands*,
and the river plains and valleys which part their
branches from one another.

(1.) The northernmost of these *chalk ranges* approaches very nearly to the oolitic in position, in form,
and in direction, save that its course is shorter. It forms
a tableland with a steep flat-topped escarpment facing
west, and with a gradual slope to the south-east and
east, furrowed by small river valleys; and it runs in
a direction almost parallel to the oolitic uplands
from the English Channel to the North Sea. Beginning on the shores of the Channel east of Lyme in
the *Dorset Heights*, it so closely adjoins the oolitic
range as to form with it for some distance almost
one tract of uplands. The Dorset Heights as they pass
northwards widen to *Salisbury Plain*, a barren and
woodless tract of undulating country covered with a
short thin grass and difficult of cultivation save in
the valleys of the streams crossing it. North of

Salisbury Plain lie the *Marlborough Downs*, whence
the range bends to the eastward, and forms the
escarpment of the *White Horse Hill* and *Ilsley
Downs*. It is now sharply broken across by a
cleft through which the river Thames passes ; and
on the further side of this gap it takes the name
of the *Chiltern Hills*, a high range of chalk downs
running to the north-east. From the Chilterns to
Hunstanton Cliff on the shores of the Wash, the
uplands are known as the *East Anglian Heights*.
Here the chalk range widens until it fills nearly
the whole of that broad tract of land which projects eastward into the North Sea ; and as its western
side becomes less steep and clearly marked, and at
the same time its eastern slopes grow more and more
gradual in their decline, it forms a wide expanse of
gently-swelling upland, entirely given up to pasture
and agriculture.

(2.) The two remaining ranges of high ground in
southern England are formed of the same chalk rocks
as the Chilterns and the East Anglian Heights—but
they differ from that range in direction, in size, and
in form.

As the two northernmost ranges, the Oolitic and the
Chalk, are bound together at their first beginning in
the south, so the two southernmost ranges, the North
and South Downs, are in like manner linked together
for a time. From the eastern side of Salisbury Plain
a broad tract of chalk uplands, known as the *Hampshire Downs*, extends in an easterly direction, and
presently breaks into two short branches, which form

the two last fingers of the hand we have imagined. These two branches, the North and South Downs, have for their starting point the country about the town of Winchester, whence the northern branch, the *North Downs*, runs due east to the sea in a narrow ridge of hills, which broadens at its eastern extremity till it ends in the North and South Forelands and the cliffs of Dover and Folkestone.

(3.) The range of the *South Downs* stretches from the same point, and has a south-easterly direction to Beachy Head on the English Channel. Like the North Downs it runs in low flat-topped ridges never mounting to 900 feet in height, or in lines of very gentle curves rising one behind the other with a monotony peculiar to chalk scenery.

UPLANDS.

Oolitic Range.

Cotswold Hills, Cleeve Hills   .   1,093 feet.
Edge Hills   826   „
Northampton uplands, highest
    point   735   „
York Wolds, Wilton Beacon   .   785   „
York Moors, Botton Head   .   .   1,489   „

Chalk Ranges.

(1.) Dorset Heights, Pillesden Pen .   910 feet.
    Salisbury Plain, highest point .   754   „
    Marlborough Downs, Milk Hill   967   „

Ilsley Downs . . . .   800 feet.
Chiltern Hills . . . .   900 to 300   „
East Anglian Heights . .   300 to 60   „
(2.) Hampshire Downs—
      Inkpen Beacon   972   „
      Highclere Beacon   863   „
North Downs . . . .   863 to 300   „
(3.) South Downs . . . .   880 to 100   „

The Plains.—By *plains* we mean the low country lying between the mountains or uplands which we have already surveyed. As we have seen, the plains are often but little lower than the uplands, and are varied in character, being sometimes broken by low hills and undulating ground, while at other times they are perfectly level. In most cases they form the valleys of rivers thrown down from the higher ground by which the plains are shut in, and are thus the most fruitful districts in England. (See p. 59).

I. The most extensive plain in England is that which lies to the west and north-west of the Oolitic Range, parting these uplands for their northern half from the Pennine Chain, and for their southern half from the group of the Welsh mountains. The two ends of this great plain lie on opposite coasts of England, one opening on the North Sea to the north of the York Moors, the other on the Irish Sea between the Welsh and Cumbrian mountains ; while its central part is sharply curved round so as to inclose the southern extremity of the Pennine Chain in the Peak. Thus the whole plain has the form of the letter **J**.

This plain is divided by its river valleys into three separate parts.

(*a.*) One portion runs due north and south, parting the Pennine Moors on its western side from the York Moors and Wolds on the east. This portion is called the **Plain of York,** or the **Valley of the River Ouse**, and consists of a broad and perfectly level tract of country, well watered by a great number of fine rivers which flow across it in every part from the neighbouring hills and moors, and make it throughout its whole length one of the most fertile agricultural districts in England.

(*b.*) The character of the country greatly changes at the base of the Pennine Chain, where the plain sweeps westward round the Peak of Derbyshire. Here it widens considerably to form the great **Central Plain of England,** or the **Valley of the Trent,** which stretches between the basin of the Severn on the west, Rockingham Forest on the east, and the Pennine range and the Cotswolds on the north and south. This broad tract of low country is broken by undulating ground, by low uplands such as that of *Cannock Chase* to the north-west, and even by steep hills thrown up abruptly from the level country, as the *Dudley Hills* near the Welsh border. It differs from the York Plain not only in its scenery but in its industries. While in some parts the country is purely agricultural, in others the industry wholly depends on mineral wealth and manufactures. Especially round the upthrow of the Dudley

Hills, beds of coal and iron-mines have been dis-
covered, which have become the centre of one of
the busiest manufacturing districts of England and
have converted the country into a vast series of
furnaces, manufactories, coal-pits, iron-works, and
potteries, whose smoke almost destroys vegetation,
and has given to this region the name of the *Black
Country*.

(*c.*) Immediately to the north-west of the manufacturing district lies that portion of the great plain which
opens on the Irish Sea, the greater part of which is
known as the **Plain of Cheshire.** It extends for
some distance northwards along the western slopes of the
South Pennine Moors, and forms a deep and wide basin
parting them from the Welsh mountains. The ground
is for the most part perfectly level, only varied by one
line of low hills, and wholly given up to pasture for
cattle, and great dairy farms. In the centre of the plain,
however, along the valley of the *Weaver*, there is a
considerable sinking of the ground, marked by brine-
springs and deep mines of rock salt, where the chief
salt manufacture of England is carried on.

The whole of this immense plain, stretching as it
does from the North Sea to the Irish Sea, is thus as
varied in character as the upland range which bounds
its eastern side. While the line of uplands is high
at either end and lower in the centre, the extremities
of the plain are level and its central part raised.
As this higher ground adjoins the lowest part of
the upland range, that is, the low heights of the Northampton uplands, it follows that the difference in

elevation between the two is not very clearly marked ; and thus the whole central district of England, from the Welsh border to the shores of the Wash, presents the same appearance of very varied though quiet, undulating country, unmarked by violent features of any kind, affording throughout its greater part excellent soil for tillage and pasture, well-watered by rivers, and rich in woodland.

II. The second valley or plain lies between the ranges of the Oolitic and Chalk Uplands. It forms a long belt of low tumbled country running in a north-easterly direction, very narrow for the greater part, but increasing to considerable width as it approaches the shores of the Wash. In the south-east the **Upper Valley of the Thames** parts the Marlborough Downs from the Cotswolds and the Edge Hills, and includes to westward the *Vale of White Horse*, and to eastward the wider and more broken *Vale of Ayles-bury*, running below the steep ridge of the Chiltern Hills. This last vale leads into the **Valley of the Ouse,** which separates the East Anglian uplands from the low undulations of Rockingham Forest. Near the shores of the Wash this tract of low country becomes still broader ; it opens in fact into a wide stretch of level, monotonous marsh, traversed by a multitude of sluggish streams. This district of the **Fens,** sometimes known as *Holland* or the hollow land, which in former times was merely a swampy wilderness full of unhealthy vapours, has been drained at immense cost of money and labour, and been made very fertile and moderately healthy, but it retains its dull monotony of scenery, long stagnant water-courses winding through flat fields, and lines of straight canals bordered by endless rows of willows.

III. A third plain parts the two northernmost ranges of chalk uplands, and constitutes one of the largest and most fertile valleys in England, and the most important in its history. It is formed by the lower **Basin of the River Thames,** the first and greatest highway of communication between England and the Continent. The valley consists of a sunken clay soil, diversified here and there by low rising grounds, through the midst of which the river runs with innumerable windings, gathering to itself on its way the many streams which fall down the gentle slopes of the hills to north and south of it. At first narrow where it is shut in between the Marlborough Downs and the Hampshire Downs, the valley widens greatly at its eastern opening, becoming more broken and undulating where it parts the Chiltern Hills and the East Anglian uplands from the North Downs. The low heights of *Epping* and *Hainault Forests*, which still retain remains of the ancient woods that once clothed them, occupy the country to the north of the Thames estuary, till they die down into the flat, swampy marshes which lie along the sea-shore. The whole of this low-lying plain is exceedingly fertile and highly cultivated, but its great importance lies in the broad estuary of the Thames as it widens out to the German Ocean, and there opens the way by which vessels pass into the port of LONDON, the capital of England.

IV. A fourth plain is that inclosed between the

steep escarpments of the North and South Downs,
and called the **Weald**. This is wholly different
from the valley of the Thames. In form indeed
it is not very unlike it, having also a wide opening
on the sea to the east, and narrowing greatly to its
opposite extremity. Its soil also consists for the
most part of heavy clay. But along its centre lies,
not a low river valley, but a ridge of sandy heights,
called the *Forest Ridge* or the *Wealden Heights*, which
forms the water-parting of the district. The rivers
of the Weald thus flow north and south from its
central heights, cutting right through the soft chalk
ranges on either side, instead of flowing like those of
the Thames basin from the hills on its borders to the
river valley in its centre. In appearance and fertility
the Weald is also very different from the Thames
valley; on all sides we see lines of low rises which were
once a mass of tangled forest, and are still covered in
great part by trees, and but partially brought under
cultivation. It contains no large towns, and its villages are supported by agriculture and the employment
given by the extensive woodland. The whole of this
Wealden district and the neighbouring chalk downs
are remarkable for their exact resemblance in geological structure to the district of Boulogne on the
opposite French coast, from which they have been
broken off by the narrow strait of Dover. The
likeness between the two shores of the channel is
thus marked as clearly in the later rocks of the east
as in the old formations of the west.

V. The smallest of all the English plains is the **Plain**

of Southampton, which lies on the southern coast. It is a low basin of clay surrounded on three sides by chalk downs—on the east by the South Downs, on the north by the Hampshire Downs, and on the west by the Downs of Dorset. Even on the southern side chalk heights can be traced in the long cliff-line of Purbeck, running out sharply to the eastward, and continued across the southern half of the Isle of Wight. In the centre of this low plain is the *New Forest*, a hunting-ground of the kings of England 800 years ago, and still a wide tract of woodland. This valley differs from the Weald in its level and monotonous character, and also in containing the two great sea-ports of the southern coast, *Southampton* and *Portsmouth*; while at the same time it differs from the basin of the Thames in being very much less fertile, and in having no large river to form a highway from its harbours to the heart of the country.

We thus see how great a part of England consists of low plains, all of which are crossed by innumerable roads, by canals measuring together 2,000 miles in length, and by railroads, which carry the produce of the country easily and rapidly to the great towns and the trading-ports. Their most important feature, however, is the system of rivers by which they are traversed, and which we must now pass on to survey.

Importance of Rivers.—As in tracing the courses of the hill masses and uplands we are led to examine the plains which lie between them, so in studying these plains, we are drawn to trace the courses of the rivers which run through their midst. This study is also forced on us by the fact that the well-being of any country depends in the first instance on its river system. There are vast tracts of lands in various parts of the world wholly forsaken and desolate because they have no rivers: all the great cities of ancient times, and by far the greater number of modern towns have sprung up on the banks of some great stream: the chief centres of population and wealth have always been found in river valleys. The reasons of this are simple. *Agriculture* is only possible where the land is drained and watered by rivers ; a supply of running water is of the first importance in *manufacturing* districts ; the *internal commerce* of a country, which is so much more important to its wealth than the external, is in its first beginnings

to a very great extent dependent on water-communi-
cation between town and town : and *foreign trade* is
developed by means of those great rivers whose
channels open a highway between nation and nation,
and whose estuaries give harbourage for ships.

The English Rivers.—England is remarkable
for the perfection of its river-system.   Other countries
have higher mountains, broader plains, a richer soil,
and a happier climate, but, owing to the amount of its
rainfall and the undulating character of its ground,
no country is more completely furnished with a network of streams, great and small, spread over its
entire surface.   The greater rivers form harbours
and waterways leading to the very heart of the
kingdom ; the lesser streams feed canals by which
goods can be easily carried from place to place ;
all alike gather together and convey across the plains
the waters needed for the fruits of the field, or for
the supply of towns.

The Two Water-partings.—This great river
system is formed into a very beautiful and perfect order, an order easy to understand when we have once
clearly grasped the lie of the mountains and uplands
by which the waters are thrown down to the plains
and there pent in till they find an outlet to the ocean.
The first great natural division of the river groups
is made by the hills and rising grounds which lie
between the sources of rivers flowing to different seas,
and so form the *water-parting*.   In England there are
two distinct lines of water-parting.   The main line
has been described in Chapter II : it passes in a

sinuous course from the Cheviot Hills along the crest of
the Pennine range, across the rising grounds of central

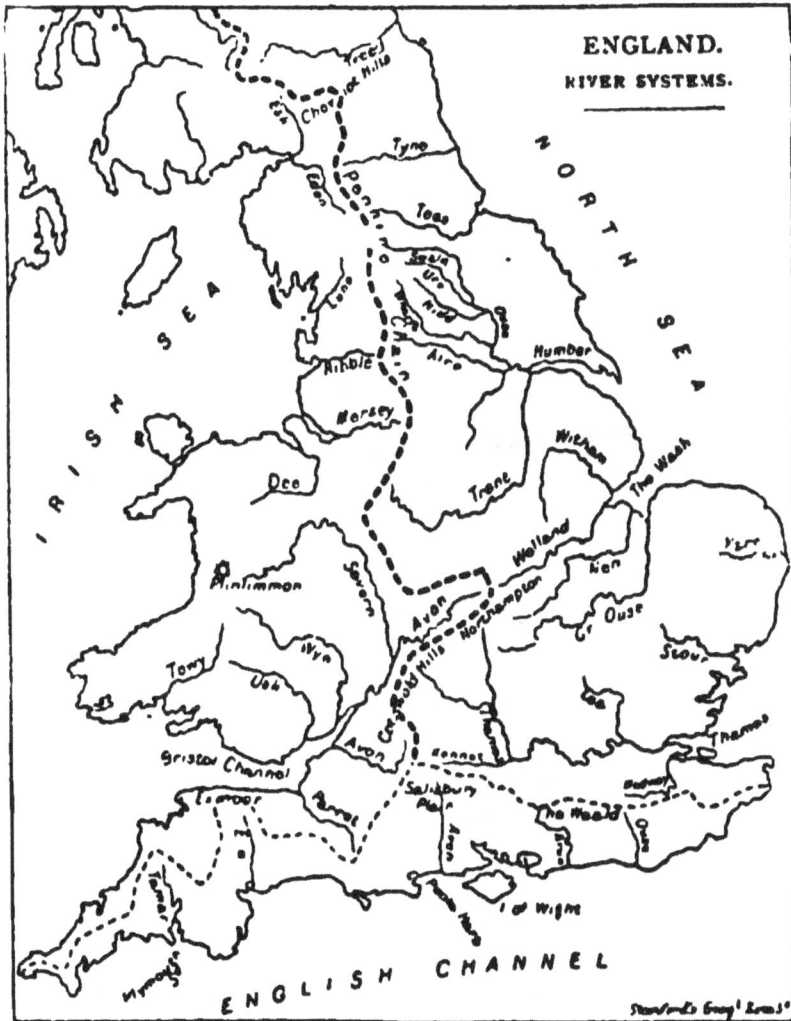

England, where it turns eastward to the Northampton
uplands, and then curves back again to end on the

Cotswolds (p. 21). This water-parting divides between the rivers which flow to the Irish and the North Seas. A second water-parting runs from east to west along the uplands of southern England, forming with the first the figure of an inverted ⊥. Its course is very irregular, passing along the Wealden Heights, the Hampshire Downs, Salisbury Plain, Exmoor, Dartmoor, and the Cornish Highlands. The streams thrown down from it flow southward to the English Channel, or northwards to the Bristol Channel and the Thames estuary.

By these two lines of water-parting England is divided into *three distinct river systems :* (1) That of the eastern country between the main water-parting and the North Sea ; (2) that of the southern slopes towards the English Channel; (3) that of the western country lying towards the Irish Sea. In this last group are included the rivers of Wales, as they pour their waters into the same sea.

Contrast of River Basins.—A glance at the map will shew the differences that exist between the rivers of these three watersheds of England. (1.) As the high ground of England strikes down its western districts, it follows that the main water-parting, where the sources of the great rivers take their rise, must lie much nearer the western than the eastern coast. From this it follows that the *greatest number of rivers flow eastward* to the North Sea. In fact, so general is this easterly direction given to the river-courses by the structure of the ground, that the four chief streams which fall into the western sea, the Dee, Severn, Wye, and Usk, all travel for the greater

part of their length from west to east, and are
only turned westward just before entering the sea.
The rivers which have a general westerly direction
are thus very few in number. (2.) For the same
reason the *eastern rivers have a long course*, and so
gather a great volume of water before reaching the
sea, while the western rivers have but a short way
to travel and are of little consequence. Indeed
the only great rivers of the west are those which
come in a winding easterly course from the Welsh
Mountains, not from the English water-parting.
(3.) The rivers of the east, as they wind among the upland ranges, receive from them many tributary streams,
so that *their basin*, or the extent of land drained by
them, *is very great*. This is not the case in western
England, where the streams fall down the mountain
slopes by the shortest course to the sea, and their
affluents are generally few and unimportant (p. 59).

The rivers of the south have a very short course,
and are all small and unimportant.

The Grouping of Rivers.—In tracing the
sources of the rivers along the water-parting we find
that they never occur singly, but always in groups of
several springs, and that these groups do not lie
evenly distributed, but are scattered unequally here
and there. For example, six of these chief knots of
head-waters are strewn along the Pennine Chain : two
more lie in the pasture-lands and low hills of the
central plain ; and three in the uplands of the oolitic
range. Sometimes these knots of springs lie wholly on
one side of the water-parting, and all the streams then

flow to the same sea; but generally the group is divided and the rivers flow in opposite directions.

As there is a grouping together of the head waters of rivers, so there is also a grouping of their *openings* into the sea. There are certain bays and estuaries lying round the coast of England which receive the greater part of the running waters of the island, for example the three great estuaries of the *Humber*, the *Wash*, and the *Thames*, into which nearly all the streams of the east coast are gathered. In considering the rivers, therefore, we shall divide them into those groups into which they are thrown by the natural formation of the country.

[A.] THE RIVERS OF THE EAST. These rivers, the most important for their numbers, their length, and the size of their basins, are divided into four groups.

I. The **Northumbrian Rivers,** or those rivers which fall into the sea to the north of the Humber, form the first group.

The river which lies to the extreme north of England, the river *Tweed*, we may pass by for the present, as it is not properly speaking an English river. The greater part of its course, in fact, lies in Scotland; it is only as it passes to the east of the Cheviot Hills and nears the sea that it touches English soil, and thence becomes the boundary line between Scotland and England.

The three rivers which fall into the North Sea below the Tweed—the *Tyne*, the *Wear*, and the *Tees*,—are closely connected in their sources, in their river-valleys, and in their industrial value. The

WATERSHED
OF THE
NORTH SEA.

springs of all three lie close together, high up in
the bleak moors of the north Pennine Range,

G

whence their course to the sea is too short to
allow them to attain very great size. But their
position gives them considerable importance. As the
Tyne and the *Wear* reach the foot of the moorland
slopes they enter at once on the great mining district
of the north, where a wide bed of coal stretches
along the sea-shore for a very great distance, and
alone yields twice as much coal as the whole of
France, and where not only the coal, but the iron,
zinc, and lead found in the rocks have given rise to
a large manufacturing industry. The Tyne passes
across the very centre of this busy region, the Wear
runs along its southern borders, and both rivers as
they enter the ocean break the level coast line, and
form harbours which are the only outlets by sea for
this immense trading industry. There is no port on
the whole continent into which so large a number of
vessels enters yearly as the port of Newcastle at the
mouth of the Tyne, and Sunderland on the Wear is
only second to it.

The history of the *Tees* is somewhat different. Its
upper course lies indeed like that of the two last
rivers among wild mountain moors, but it finally
emerges on low ground at the extremity of the
agricultural plain of York, where the only manufactures are those of sail-cloth and shipping for the
northern trading ports. At its estuary, however, the
Tees lies just under the great iron district of the
Cleveland Hills, and it thus shares in the importance of the other rivers of this group in forming
a sea-port for the commerce of northern England.

II. Passing south we come to the great group of
rivers which are gathered together into the **Humber**
(p. 81). The Humber is an estuary whose channel is
out through the range of uplands which parts the
plain of York from the North Sea : to the north of it
lie the Wolds of York, to the south the Wolds of
Lincoln. As this is the only opening in the long
barrier of uplands which runs along this portion of
the coast, and as it leads to the very heart of the great
plain shut in between these uplands and the Pennine
range, it has from the mere fact of its position become
the channel into which all the rivers watering this district are poured, and by which they are carried to the
ocean. Here both the rivers of the plain of York,
severed from the sea by the northern part of the oolitic
uplands, and the rivers of the central plain of England,
equally severed from the sea by the central part of
the same range, find their natural outlet (see p. 59).
This Humber river-group is therefore the largest in
England, draining as it does such an immense extent
of country, not only the York Plain but also the
great plain of Central England, an area in all of 9,500
square miles. By far the greater number of its rivers
have their head-waters in the moors of the Pennine
range, but as they fall to the plains they are taken up
by two main streams. The *Ouse* gathers into itself
all the waters from the northern half of the plain ;
the *Trent*, all those from the southern. Thus the
rivers which unite in the Humber are naturally divided
into two distinct groups, (1) those brought to it by the
Ouse, and (2) those brought to it by the Trent.

(1.) The **Ouse and its tributaries** water the rich plain of York, which lies between the Pennine range and the Yorkshire Moors. To trace the sources of the Ouse, we must go up into the Pennine moorlands, to a point where two streams, the *Swale* and the *Ure*, take their rise side by side and flow in deep valleys down the mountain slopes to the low land, where both turn southward and soon after unite to form the Ouse. As this river winds through the rich plain in a long course of 114 miles it presently receives one after another three more streams coming from the same region whence flowed its first two tributaries. These three rivers, the *Nidd*, the *Wharfe*, and the *Aire*, are all thrown down the slopes of a high range of moorland which forms the southern wall of the upper valley of the Ure, and from their wild valleys among the hills they enter the plain one beyond the other, and so fall into the Ouse at equal distances, forming by their courses three parallel curves. The first stream, the Nidd, has the shortest course; then comes the Wharfe; while the outermost circle of the Aire is the greatest of all. Thus five rivers flow into the Ouse on its *western* banks, and these are all drawn from the same district in the Pennine range. On its *eastern* banks the river takes up only one stream, the *Derwent*, which is thrown down from the Yorkshire Moors. The Derwent rises among the heights close to the coast; from these it falls down to the broad Vale of Pickering, and thence flows into the Plain of York, where after some time it joins the Ouse, and thus passes on to the Humber and to the sea.

One river yet remains to be added to this group. The river *Don* rises far from the rest in a distinct group of head-waters lying near the southern end of the Pennine Chain, and flows in a north-easterly direction to join the Humber almost at the point where the Ouse empties itself into that estuary. Its course from the moorlands, and the place of its junction with the Humber, mark it out as belonging to the group of the Ouse rivers, whose number is thus made up to seven.

These rivers of the Pennine Moors have some resemblance to those of the Northumbrian group. Their wild mountain valleys are of the same character, and the two southernmost of the streams as they descend the hill slopes also enter on a great coal field, that which lies stretched along the eastern side of the Pennine Chain from the valley of the Aire as far as the valley of the Trent or the central plain of England. The Aire, which marks roughly speaking the northern limit of this coal-field, passes across its broadest and busiest part, while the Don cuts through its centre, where it is narrower. But as this mining district is bounded eastward, not by the sea, but by a level inland plain, the rivers do not play the same important part as they do in the north in forming outlets for foreign trade. The main value of the whole group of streams, as they are gathered in from east and west by the Ouse, lies partly in their supplying the running waters needed for manufacturing purposes in the coal-field, partly in their course through the plain, which they water and make fertile in every part;

while both plain and coal-field are opened to trade with foreign lands by means of the Humber.

(2.) As the Ouse and its tributary streams occupy the north-eastern part of the vast plain which half encircles the Pennine Moors, so **the Trent and its tributaries** drain the whole of its central and south-eastern portion.   The *Trent* rises on the south-western slopes of the Pennines, on the confines of the low plain of Cheshire.   It first flows downwards to the southern extremity of the moorland range, sweeps right round it as it traverses the low broken country of central England, then bends northwards till its course forms a great semicircle, and empties itself at last into the Humber after a journey of 147 miles.   It receives four rivers on its way to the sea, two on the right hand, and two on the left, all of which meet it in that part of its course which lies exactly below the extremity of the Pennines at the Peak; and from this point it widens so much as to become navigable, and thus opens a waterway from the very centre of England to the sea.   Its two northern tributaries, the *Dove* and the *Derwent*, descend from the moors of the Peak in parallel channels not very far apart; the Derwent from its source in the heights close to the Don forming a narrow river valley of extreme beauty among the steep rocks of the mountain range.   On the south the Trent receives the *Tame* coming from Cannock Chase, and the *Soar*, which springs in a rising ground more to the west, and empties itself into the Trent just opposite the mouth of the Derwent.   By these two rivers it drains that irregular tract of country which lies between the

hills bounding the Severn valley, and the upland
range of Rockingham Forest (see p. 59)., The group
of rivers carried to the Humber by the Trent is
thus smaller in number than the group gathered
up by the Ouse, but this inferiority in numbers is
fully atoned for by the great size and importance of
the main stream, by the length of its course across
almost the whole width of England, by the immense
area of fertile country which it drains, and by the
extent of its navigable waters.

These two great river-groups in their junction with
the Humber unite in one most important work, the
bringing the whole of the agricultural and manufacturing industries of central and north-eastern England
into connection with the greatest seaport of the north-
east, the port of *Hull* on the Humber. The Ouse
and its affluents flow through the agricultural
district of the plain of York, and among the
coal-mines and iron-works of the Pennine slopes, and
the woollen manufactures of the moors. The Trent
and its tributaries pass in their course by a great
number of the chief manufacturing industries of
England—its potteries, its iron-works, its coal-mines,
its breweries, its manufactures of lace and stockings, and the centres of its corn and cattle trade.
The immense extent and variety of the industries thus
represented by these two river-groups as they fall
together into the Humber give to that estuary its
very great importance as a sea-port. Besides its position as the outlet of all these rivers, the Humber
has a further value derived from its position relatively

to other harbours and to the Continent; it is
far removed from any rival port, the nearest being
that of the Thames estuary in the south-east, and it
thus forms the sole highway for trade to North Europe
and the ports of the Baltic.

III. A third river-group, which lies next to the
south, is formed by the streams that unite in the
estuary of the **Wash**, as the rivers of the Ouse and the
Trent unite in the Humber. *The Wash* is a wide,
shallow inlet of the sea, with an area of about 300
square miles, bounded on its land side by a low
swamp, and utterly useless for all purposes of
harbourage. Its basin, 5,800 square miles in extent,
or more than one-ninth part of England, is parted from
that of the Trent by the low uplands of Northampton
and Lincoln (see p. 59). The direction of this higher
ground is marked by the springs of the four streams
which are thrown down its gentle slopes and emerge
on the broad, dreary flats of the Fens, where their
waters once mingled in a vast marsh, and where even
now they have no clearly defined valleys, but wind
sluggishly in constantly changing courses through
tracts of level ground reclaimed by immense labour,
or through low, unhealthy swamps, to the desolate
shores of the Wash (see p. 81).

The *Witham* rises in the Lincoln uplands and flows
northwards along their western side for some distance,
then suddenly bends east, cuts through a depression
in the ridge, and turns south on its other side, passing
through the swamps inclosed between the oolitic
uplands and the chalk Wolds to the shores of the

Wash. The two next rivers, the *Welland* and the
Nen, take their rise among a group of river sources in
the Northampton uplands. They both flow to the north-
east not far apart from one another, inclosing between
them the high ground of Rockingham Forest, and
drawing closer together as they sink into the low Fen
country near the Wash. The last river of this group,
the *Great Ouse*, rises yet further south, on the borders
of the Northampton uplands, and soon flows into the
low, irregular ground which parts the oolitic from the
chalk ranges. Here after endless windings and turnings
in the tumbled ground it bends sharply towards the
East Anglian uplands, and flows in a direct line to the
Wash beneath their white chalk cliffs, with the wide
expanse of the Fens stretching away in the distance
on its opposite banks. The Ouse is very much the
greatest of all these streams, being 156 miles long,
or nearly twice as long as the Witham ; and its basin
extends under almost the whole length of the East
Anglian Heights and the Chiltern Hills.

These rivers of the Wash form a striking contrast to
the group of the Humber. They rise on that part of
the water-parting which most nearly approaches the
eastern coast, and hence their courses are generally
shorter. As they all have their springs on the same
low uplands, and as all flow through the same monotonous level to the same harbourless estuary, the
character of the country they traverse is everywhere
alike. It is a purely agricultural district, thinly
inhabited, with scarcely a large town, and lying entirely
apart from the manufacturing industry of the Humber

rivers. For commercial purposes their waters are all but useless.

IV. A fourth group is composed of the **Rivers of the Basin of the Thames and its Estuary.**

BASIN OF THE THAMES.

The whole importance of this group depends on the *Thames* itself. Rising little more than 300 feet above the sea-level, it is the largest river in England, and is navigable for nearly the whole of its great length of 215 miles. It thus forms a water-way across southern England in the same way that

the Trent does across central England ; and this water-
way communicates with the sea by a wide estuary
which is the most important of all English harbours.
The basin of the Thames has an area of 6,000
square miles, and may be divided into two parts.
(a.) From its springs in the Cotswold Hills, not
far from the estuary of the Severn, the river flows
through the low ground lying between the Edge Hills
and the Marlborough Downs, and thus severs the
oolitic from the chalk uplands (see p. 59). After a
time it turns due south and cuts straight through the
chalk escarpment which before bounded its valley,
by a channel which parts the Ilsley Downs from the
Chilterns. (b.) It now enters on the second part of its
course, and from this point its valley consists chiefly of
a great basin of low clay inclosed between ranges of
hills, from which lesser streams are thrown down ; to
the north lie the Chilterns and the East Anglian uplands, to the south the Hampshire Downs and the
Wealden Heights. The scenery of the Thames valley
is throughout of a very quiet character. The only
high ground is at the point where the river cuts
through the chalk escarpment; everywhere else the
river winds peacefully between level banks through a
rich agricultural country parted by the whole width
of two upland ranges from the manufacturing districts
of central England. But as it widens to its estuary
the scene suddenly changes. For this estuary, the
most splendid natural harbour of the British Islands,
has by its form and position been destined to become
the trading centre of the world. It is wider and

more commodious than the Humber, deeper and safer than the Wash. Placed in the south-eastern angle of England, it opens on the narrow channel which forms a highway for the commerce of northern Europe, while lying as it does midway between all the European seas, from Gibraltar to the North Cape, and being situated at the centre of the main lines of navigation of the world, it has necessarily drawn to itself commerce from every quarter of the globe. On its banks has grown up the greatest and most populous city of the earth,—*London;* and from this city to the sea, throughout the entire length of the estuary, its shore is lined with warehouses and dockyards filled with shipping from foreign lands, while its waters are covered with countless vessels, carrying the immense trade of the greatest of all trading countries.

Four streams join the Thames on its northern banks, and four on its southern. Of the northern streams two, the *Cherwell* and the *Thame*, meet it in the first part of its course, just before it breaks through the chalk uplands; two more, the *Colne* and the *Lea*, join it in its lower valley. On the southern side falls in the *Kennet*, whose course lies below the Hampshire Downs, and which joins the Thames at the point where it passes into its lower valley. Further on two rivers are thrown down from the south, the *Wey* and the *Mole;* and in the east, where the Thames estuary opens into the ocean, it takes up the *Medway*, which alone is of any considerable size and navigable for some distance, and which by a branch called the *East Swale* encircles the island of Sheppey at its mouth.

These rivers flow directly into the Thames; but there are besides a few small streams included in the same group which fall into the wide bay formed by its estuary. Most of these come from the low eastern slopes of the East Anglian uplands. The *Yare* and the *Waveney* empty themselves into the North Sea as the coast bends inwards towards the Thames estuary (see p. 81); farther south are the *Orwell* and the *Stour*, which unite at their mouths to form a small estuary. On the southern shores of the bay a second river *Stour* falls into the North Sea, which once enclosed by a branch stream, now dried up, the Isle of Thanet at its mouth. These rivers are of no importance in their very short course, save for the little harbours formed at their mouths.

[B.] THE RIVERS OF THE SOUTH. All the rivers of southern England have their sources along the lesser line of water-parting which crosses the island from the Atlantic to the North Sea, and flow northwards to the Thames and Bristol Channel, or southward to the English Channel (see p. 77).

I. The first group is formed by the **rivers of the Weald**, rivers small in size, crossing a country but thinly inhabited, and of no commercial importance. All have the same origin in a line of high ground which crosses the centre of the Weald from east to west (see p. 90). This line of heights throws down small streams to north and south. The first cut channels for themselves through the North Downs, and so make their

way into the Thames or the German Ocean. Among
these are the *Mole*, *Medway*, and *Stour*. The rivers
of the southern slopes are of very little consequence.
The largest of them is the *Ouse*, which falls into the
English Channel west of Beachy Head: and yet
further to the west is the *Arun*.

II. A second group, yet smaller in numbers but
made up of larger streams, is that of the **rivers of the
southern chalk downs** (see p. 96). The plain of
Southampton and of the New Forest is inclosed on
three sides by chalk uplands, whose waters are thrown
down to the English Channel. The smallest of these,
the *Itchen*, and the *Test* or *Anton*, on the eastern side
of the New Forest, fall into the estuary called Southampton Water, behind the Isle of Wight. To the
west of the New Forest is a much larger river, the
Avon, navigable for many miles. A fourth river, the
Frome, flows in a valley running east and west below
the long line of the Purbeck heights, and at its mouth
forms the inlet of *Poole Harbour*, whose opening lies
in a right line with the Isle of Wight.

III. The **Rivers of Devon and Cornwall** form
a third group, also of small importance, since in this
narrow peninsula the distance from the heights where
the rivers spring to the sea is so short that no streams
of any size can be formed (see p. 96). The two largest
on the south-eastern side are the *Exe* and the *Tamar*.
The Exe rises in the bleak Exmoor heights and flows
southwards across the plain below them to the
English Channel. The Tamar has its source in the
heights above Bideford Bay whence it also flows south

to form at its mouth on the Channel the estuary of
Plymouth Sound. Three streams are thrown down
from the water-parting to the north-west : the *Tawe*
and the *Torridge* to Bideford Bay, and the *Parret*, to
the Bristol Channel.

[C.] THE RIVERS OF THE WEST. When we pass
from the southern to the western coast of England
we find a water-system of far greater importance,
containing groups of rivers of very great size and of
even greater commercial importance than the rivers
of the eastern plains. It is true that the streams
which flow westward from the main water-parting
of England are fewer and generally shorter than
those which flow eastward. But among the rivers
of the western coast we must include the streams
drawn from the mountains of Wales, of which some
flow eastward into English land and unite with streams
from the English water-parting to form rivers of considerable size. The main cause, however, of the
commercial value of these streams lies in the great
advantage which they have over the eastern rivers
in their position to the westward, by which they
command the whole of the Irish trade, and the
vast and constantly increasing trade with the New
World.

The rivers of the west are divided into four distinct
groups. Two of these groups, the *rivers of the Bristol
Channel*, and the *rivers of the plains of Cheshire and
Lancashire* are great centres of commercial activity
corresponding to the groups of the Thames and the
Humber in the east (see p. 77). The two remaining

groups, those of the *Welsh* and the *Cumbrian mountains,* are composed of small rivers of less note.

I. In the **group of the Bristol Channel** the rivers of England and Wales are so closely united

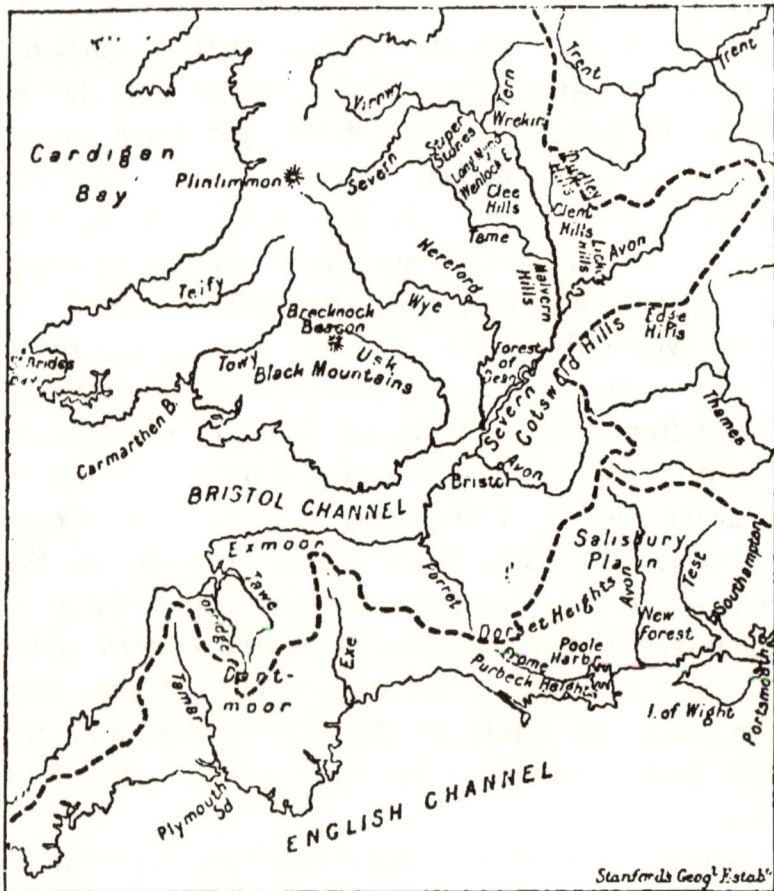

WATERSHED OF THE BRISTOL CHANNEL.

that it is difficult to separate them; for while some streams of this group belong wholly to England, others belong equally to both countries.

Five chief rivers fall into the inner part of the Bristol Channel from the various heights—Welsh mountains or English uplands—which lie to north and south and east of it. The greatest of all these rivers is that which lies in the centre of the group —the *Severn;* into its estuary the four other rivers pour their waters, two on the northern, and two on the southern side.

(1.) The *Severn* takes its rise near the summit of the great mountain of Plinlimmon in the centre of Wales. Its upper course lies in Wales, where it flows in a north-easterly direction through the wild moorlands which branch off from Plinlimmon; but as it breaks out on the plain which bounds the Welsh mountains it enters on English ground (p. 59). Here the river bends suddenly southwards and skirts the borders of the irregular heights of eastern Wales. On its right bank in this portion of its course lie long ridges of slate-hills, outliers of the mountains beyond, the Stiper Stones, the Long Mynd, Wenlock Edge, and the group of the Clee Hills; along its left bank stretches the plain of central England broken by the Wrekin and the Dudley Hills. The rest of its course to the Bristol Channel lies in a deep valley not many miles wide, inclosed between steep ridges of hills. The western wall of this valley is formed by the Abberley and the Malvern Hills, the east ern by the Clent and the Lickey Hills; while the broader opening which forms the estuary of the river lies between the Cotswold Hills and the Forest of Dean.

The Severn has a total length of 158 miles, and is thus the third greatest river in England; and its basin is not only of considerable extent—4,350 square miles —but also of great commercial importance, since its estuary is driven through the midst of the coal-measures of the south-west, while its affluents lead into the very centre of England, connecting it with the basins of the Trent and the Thames. By the great length of its navigable course from the point where the river reaches the plain, and by the splendid estuary at its mouth, which is flooded at high water by a tide that rises higher than in any other harbour in Europe and carries large vessels safely over shoals and sandbanks, the Severn is naturally fitted to form one of the great trading centres of the west; and its port of *Bristol* has drawn to itself the chief commerce of Ireland and the West Indies.

The tributaries of the Severn are mostly small. It receives two streams from the Welsh mountains: one, the *Virnwy*, reaches it as it first touches the plain; the second, the *Teme*, makes its way to join it through the gap between the Abberley and the Malvern Hills. Two small streams come to it from the rising grounds of the central plain of England, the *Tern* from the north, the *Stour* from the Dudley Hills on the east. But a much greater river than these empties itself into the Severn; this is the *Avon*. For its source we must go far away to the upland range near Rockingham Forest where we have already found the springs of the Welland and the Nen (see p. 81). While these rivers flow eastward to the Wash, the Avon falls down the western slopes of the uplands and winds along the base of the oolitic range past the Edge Hills to the Vale of Evesham below the Cotswolds, and so on to the Severn.

(2.) The second of the rivers which merge in the Bristol Channel is the *Wye*. At the opening and close of its career the Wye is closely connected with the Severn : their springs lie side by side on the same mountain of Plinlimmon, and they empty themselves side by side into the same estuary. But through the rest of their courses they are widely separated. While the Severn turns to the north-east towards the plains of Cheshire and central England, the Wye flows south-wards, and journeys among the desert, almost unin-habited moorlands of south-eastern Wales, till it enters from the west on the fertile plain of Hereford, and winds among its hills, its orchards, its hop-gardens, and its cider manufactories, to the Forest of Dean. From this point it soon makes its way to the Bristol Channel.

(3.) The *Usk*, the third river of the group which meet in the same channel, is a much smaller stream than the Wye, though it is more striking for the grandeur of its scenery. This river rises at the western end of the Black Mountains, and flows in a deep valley eastward along the foot of the chief mountain chain of South Wales, the Black Mountains and Brecknock Beacon, whose rugged sides tower above the deep bed of the stream like a gigantic wall. It finally bends round the eastern extremity of this chain, and falls into the Severn estuary not far from the Wye.

(4.) The two streams which flow into the southern side of the Bristol Channel, the *Avon* and the *Parret*, come from the uplands which occupy the country to the east of it. The Avon rises in the Cotswolds, and though its course is short, at its opening into the estuary of the Severn it forms the most important harbour of the Bristol Channel, that of Bristol itself. The Parret comes from the Dorset heights, uplands of the southern water-parting. Its lower valley lies between the two ridges of the Polden and Quantock Hills as far as Bridgwater Bay, above which lies the port of Bridgwater with a considerable shipping-trade.

II. A second group of western waters is formed by the **Welsh rivers**, whose courses lie wholly within Wales. The line of high ground in which these streams take their rise strikes down the centre of Wales from north to south, near the western coast (p. 46); and its chief rivers—the *Severn*, the *Wye*, the *Usk*, and the *Dee*—are thrown down from it eastward into the deep depression which extends from the Irish Sea to the British Channel, and severs England from the Welsh mountain group. In this low ground these rivers enter on English soil, and as they lie to westward of the English water-parting, they are by the character of the country thrown into the Western Sea (see p. 21). But there are four lesser streams which never cross the borders of Wales. Two of these rivers lie in South Wales; the other two in North Wales.

(*a.*) The *Towy* and the *Teify*, the two streams of the south, have their sources close together in the upper end of that long range of heights which begins in

Plinlimmon and ends on the shores of St. Bride's Bay (see p. 96). The Towy is thrown down their southern slopes, and pours its waters into the sea in Carmarthen Bay. The Teify runs in a deep valley along the middle part of the range, and finally falls down its northern slopes into Cardigan Bay.

(*b*.) The two northern streams are the *Conway* and the *Clwyd*, which run on either side of the low slaty hills of Denbighshire to the Irish Sea (see p. 102). The little valley of the Conway, with its pleasant fertile fields, lies between the Denbighshire hills and the barren mountains of the Snowdon range. On the eastern side of the same Denbighshire hills lies the parallel valley of the Clwyd, shut in to the eastward by the hills of Flintshire.

III. A third group, that of **the Rivers of Cheshire and Lancashire,** is made up, like the group of the Bristol Channel, of streams from both England and Wales. The low marshy coast of the plain which lies along the Irish Sea between the Welsh and Cumbrian mountains is broken by three great estuaries, those of the *Dee*, the *Mersey*, and the *Ribble*, the first two lying very close together with but a narrow neck of land between them, while the third is half-way up the line of coast to the north of these.

(1.) The head waters of the *Dee* rise in the range of the Berwyn Mountains in Wales. Not far from its source the river passes through Bala Lake, and winding along the base of the mountain range escapes by the vale of Llangollen into the low plain of Cheshire, where it turns northward, and flows in a wide full stream by the

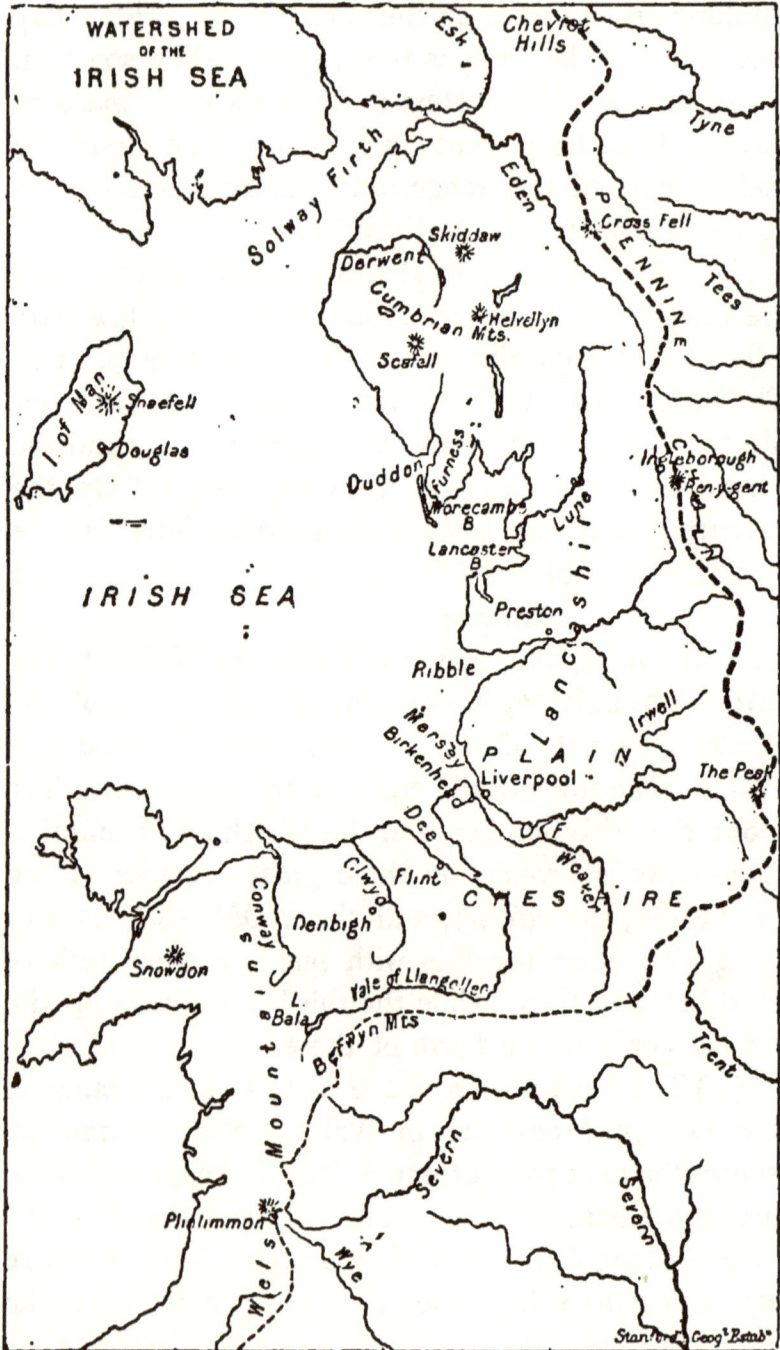

WATERSHED
OF THE
IRISH SEA

Esk

Cheviot Hills

Tyne

Solway Firth

Eden

Cross Fell

PENNINE

Tees

Skiddaw

Derwent

Cumbrian Mts.

Helvellyn

Scafell

I. of Man

Snaefell

Douglas

Duddon

Furness

Morecambe

Lancaster B

Lune

Ingleborough

Pen-y-gent

IRISH SEA

LANCASHIRE

Preston

Ribble

Mersey

Birkenhead

PLAIN

Irwell

The Peak

Liverpool

Dee

OF

Weaver

CHESHIRE

Conway

Clwyd

Flint

Denbigh

Snowdon

Montgomery

L. Bala

Vale of Llangollen

Berwyn Mts

Trent

Plinlimmon

Welsh Mts

Wye

Severn

Severn

Stanford's Geog. Estab.

eastern borders of the Welsh mountains till it falls by a very wide estuary into the sea. This estuary once formed a port of great value, but has in the course of centuries been so silted up as to make it worthless for shipping purposes, and is now at low water a mere expanse of sand.

(2.) The *Mersey* is the most important river of this group. It is formed by the union of several streams coming from the Pennine moors, and its course passes across the low plain between these moors and the sea. By the magnificent estuary through which it enters the sea, and by the marvellous wealth of the country it traverses, the Mersey has become the most important river in England after the Thames. On its southern bank lies the luxuriant pasture-land of Cheshire, a wide stretch of low country which reaches from the Mersey to the Welsh mountains (p. 59). Along its northern banks is a vast coal-field which, beginning on the high slopes of the Pennine moors, extends through southern Lancashire nearly to the sea-shore, and is occupied throughout its whole extent by the cotton manufactures of England. Easy communication with these wealthy districts on either side of the Mersey is secured by two streams which flow to it from north and south. The *Irwell* comes to meet it from the north, having first gathered up into itself all the streams which traverse the *manufacturing* district. The *Weaver*, which falls into its estuary, comes from the rich southern *pasture-lands*, and through the heart of the sunken district in the centre of the plain, where the *salt-mines* and springs of brine occur, the centre of the English salt manufacture. The Mersey, therefore, by its own course and by that of its tributaries, comes in contact with both the agricultural and the manufacturing industries of England, and opens for them a highway to the sea of the first importance. As the Humber, on the opposite coast of England, is the sea-port of the woollen trade, so is the Mersey the sea-port of the cotton trade. But the Mersey possesses great advantages over the Humber, both as being the geographical centre of Great Britain and Ireland and the natural point of convergence for their commerce, and as having its outlook, not towards Norway and Denmark, but towards America with its immense commercial resources. It has even taken a higher place than the estuary of the Severn by its situation on the very borders of those coal-measures on which the greatest manufactures of the world have grown up. There is no port in any country, save that of London, with so great a trading industry as the Mersey; and in fact, if it is below London in the amount of its imports, it exceeds it in its exports. Not less than half of the manufactured goods sent out of the British Islands pass through its great ports of *Liverpool and Birkenhead*, where its estuary is lined with docks for five miles, each dock representing a different trade and different nation.

(3.) The third river of this group, the *Ribble*, is of less importance. It rises in that knot of head-waters where the greatest number of the northern rivers of the Humber have their springs, the Wharfe, Aire, &c. Its upper course lies among barren moorlands; its lower

valley and estuary cut through the marsh which
here lines the shore. The harbour of *Preston* lies at
the head of the estuary.

IV. The northernmost river group of the western
coast is that formed by the **Cumbrian rivers.** Four
streams belong to this group. Two of these rise
near the line of the main water-parting of England,
and form by their valleys the eastern boundary of the
Cumbrian mountains ; while two smaller streams rise
within the circle of the Cumbrian mountains themselves and are thence thrown down westward to
the sea.

(*a.*) The two greater rivers are the *Eden* and the
Lune, whose springs lie close together on the Pennine
moors, on the opposite side of the water-parting from
the sources of the Ure and the Swale. As they fall
down the hill-sides to the lower ground, the Lune
turns south and flows in a steep, wild, and narrow
valley to Lancaster Bay ; while the Eden bends northwards and passes through a broad fertile plain,
bounded by the Cumbrian hills on the west and by
the Pennines on the east, to the Solway Firth. This
valley forms a valuable tract of rich agricultural land,
thrust in like a wedge between the barren hills of
northern England.

(*b.*) The two lesser streams of this group spring
from the Cumbrian heights; these are the *Duddon*,
which flows in a wild mountain valley to the sea on
the western side of Furness ; and the river *Derwent*,
which rises in the centre of the mountains, and drains
many of the lakes on its way westward to the sea.

TABLE OF RIVER GROUPS.

I. Northumbrian Rivers—
  (*Tyne, Wear, Tees.*)

II. Rivers of the Humber—
  (1.) Ouse—
    (*Ure, Swale, Nidd, Wharfe, Aire, Don, Derwent.*)

  (2.) Trent—
    (*Dove, Derwent, Tame, Soar.*)
III. Rivers of the Wash—
  (*Witham, Welland, Nen, Ouse.*)

IV. Rivers of the Thames basin—
  (*Thames, Cherwell, Thame, Colne, Lea, Kennet,
  Wey, Mole, Medway, Yare, Waveney, Stour.*)

V. Rivers of the Weald—
  (*Ouse, Arun.*)

VI. Rivers of the Southern Downs—
  (*Itchen, Test, Avon, Frome.*)

VII. Rivers of the Bristol Channel—
  (*Severn, Wye, Usk, Avon, Parret.*)

VIII. Welsh Rivers—
  (*Towy, Teify, Conway, Clwyd.*)

IX. Rivers of Cheshire and Lancashire—
  (*Dee, Mersey, Ribble.*)

X. Cumbrian Rivers—
  (*Lune, Duddon, Derwent, Eden.*)

Rivers.	Length in miles.	Area of Basin in square miles.
Thames	215	6,160
Severn	158	4,350
Humber { Trent . .	147	
Ouse . .	114	9,550
Humber .	38	
Great Ouse. . . .	156	2,980
Wye	135	1,609
Nen	100	1,077
Tees	79	708
Dee	70	813
Witham	80	1,079
Eden	69	915
Avon (of Bristol) . .	62	891
Usk	65	634
Tyne	73	1,130
Welland	70	760
Wear	65	456
Yare	48	880
Avon (of Salisbury) .	61	1,132
Mersey	68	1,000

ENGLAND.
COUNTIES.

CHAPTER VIII.

THE ENGLISH COUNTIES.

Political Divisions.—We have now traversed, however rapidly and roughly, the whole face of England; we have surveyed its mountains and uplands, its plains and river groups, till the whole of its physical structure lies before us. But the physical structure of a country is only the scaffolding on which its political geography is built up. Hill and valley and plain have become the habitation of man, and have shaped by their form and character the *political* divisions of the English nation. In very early times they furnished boundaries and settlements for the various tribes of the English race, or for their conquerors. But as England drew together into a realm, and the tribes were joined into one people, these early kingdoms or settlements died down into mere divisions, for the purposes of local government, which are known as *Counties* or *Shires*. Though we cannot dwell here on the origin of these divisions, we must remember that it has left its stamp on their form and character; and that it is only by the history of their upgrowth that we can explain the great differences of size between them, as well as other differences, such as the formation of their names, their local dialects, and the peculiar customs and character of their peoples. Here however we have to deal with counties simply as geographical divisions, and as enabling us to take a survey of the land in its political and industrial character, as we have already done in its physical structure.

England is broken up for political purposes into forty Shires, each one of which has its *County-town* where the county business is carried on, such as Quarter Sessions, or Assizes. These towns were formerly the most important places in the county, but

have in many cases been outgrown in size by other
towns in the shire.

Groups of Counties.—It is by the river system
of England, which we have already minutely studied,
that the political divisions of the country have
been mainly determined. It is only when we tho-
roughly understand the lie of the great river-basins,
and their position with regard to the heights which
limit them, that we can grasp the form and relations
of the counties which occupy their area, and which
naturally group themselves in these great geographical divisions. Thus the counties generally known
as the *Northern Counties* are those traversed by the
waters which fall to east and west from the north
Pennine moors ; while the *North-western Counties*
form the basins of the Ribble and the Mersey. The
West Midlands occupy the basin of the Severn : those
which are called the *Midland Counties* lie in the basin
of the Trent, and *Yorkshire* in that of the Ouse ; while
the *East Midlands* gather round the waters which flow
into the Wash. Another group lies along the banks of
the Thames : the *Eastern Counties* border the estuary
of the Thames : the *Southern Counties* are those
watered by the streams of the southern water-parting
which flow to the English Channel : and the *South-
western Counties* occupy the slopes of the water-
parting of Devon and Cornwall. In this manner the
forty counties into which England is divided are
gathered together into *nine* distinct groups, which are
given, with the names of the counties that compose
them, in the following table :—

I. Counties of the Northern River-basins—
　　(*Northumberland, Durham, Cumberland, West-moreland.*)
　　Area, 5,248 square miles ; population, 1,356,998.

II. Counties of the Ribble and Mersey basins—
　　(*Lancashire, Cheshire.*)
　　Area, 3,010 square miles; population, 3,380,696.

III. Counties of the Severn basin—
　　(*Shropshire, Worcestershire, Gloucestershire, War-wickshire, Herefordshire, Monmouthshire.*)
　　Area, 5,580 square miles ; population, 2,076,595.

IV. Counties of the Humber basin—
　　(*Derbyshire, Leicestershire, Staffordshire, Notting-hamshire, Yorkshire.*)
　　Area, 9,859 square miles ; population, 4,263,144.

V. Counties of the Wash—
　　(*Lincolnshire, Rutland, Northamptonshire, Hunt-ingdonshire, Bedfordshire, Cambridgeshire.*)
　　Area, 5,552 square miles ; population, 1,099,434.

VI. East Anglian Counties—
　　(*Norfolk, Suffolk.*)
　　Area, 3,597 square miles ; population, 787,525.

VII. Counties of the Thames basin—
　　(*Essex, Hertfordshire, Middlesex, Buckingham-shire, Oxfordshire, Berkshire, Surrey, Kent.*)
　　Area, 7,098 square miles ; population, 5,688,685.

VIII. Counties of the Southern Water-parting—
　　(*Sussex, Hampshire, Wiltshire, Dorsetshire.*)
　　Area, 5,471 square miles; population, 1,414,854.

IX. Counties of the South-western Water-parting—
　　(*Somersetshire, Devonshire, Cornwall.*)
　　Area, 5,590 square miles ; population, 1,427,200.

The great differences which exist between these groups of counties, in area, in structure, in scenery, in industry, and in population, depend on the character of the river basins in which they lie. Five of them, the first four and the last, are specially distinguished as comprising the mining and manufacturing districts which gather round the mountain masses of northern and western England. They form therefore the most populous, the most wealthy, the most industrious, and the most influential half of England. The following table, which gives the position and extent of the coal-measures of southern Britain, shows the counties whose industry is thus stimulated by mineral wealth. The remaining four groups of the eastern and south-eastern counties are wholly agricultural and pastoral, are less densely peopled in proportion to their extent, and contain fewer towns of importance. The district immediately round London in the lower valley of the Thames forms the only exception to this general rule.

COAL-FIELDS OF ENGLAND AND WALES.

(Coal-Fields of the Pennine Chain.)

	Area in sq. miles.
Northumberland and Durham . . .	460
York and Derby	760
Lancashire.	217
North Stafford	75
South Stafford	93
Leicestershire.	15
Warwickshire	30

(Coal-Fields of the Cumbrian Mountains.)

Whitehaven 25

(Coal-Fields of the Welsh Mountains.)

Anglesea	9
Flintshire	35
Denbighshire	47
Shropshire	28
South Wales	906
Forest of Dean	34
Bristol and Somerset	150

Total area . . . 2,884

CHAPTER IX.

THE COUNTIES OF THE NORTHERN RIVER-BASINS—OR THE NORTHUMBRIAN COUNTIES.

THE highest point of the Pennine range is the mountain of Crossfell; and from its sides or from the neighbouring heights are thrown down the four great rivers which water the north of England (see p. 77). To the east flow the Tyne, the Wear, and the Tees; to the westward a number of little tributary streams swell the waters of the Eden on its way to the north. The basins of these rivers and the surrounding country form a group of four counties, which we may term the **Northumbrian** shires, two of which lie on either side of the water-parting. If we look in fancy from Crossfell to the north and north-east, the course of the Tyne leads us to *Northumberland*, a great tract of country wedged in between the Cheviot Mountains and the North Sea. As we look eastward the deep valley of the Wear broadens into the plains of *Durham*, which extend along the shores of the North Sea as far as the valley of the Tees. To the south, on the borders of the moors of *Westmoreland*, we see the springs of the Lune and of the Eden ; as we track the Eden in its course to the north-west, we pass from Westmoreland into *Cumberland ;* while beyond the Eden valley

NORTHUMBRIAN COUNTIES.—Towns and Rivers.

we see rising the mass of the Cumbrian Mountains, overlapping the borders of both counties and shutting out from view the Irish Sea.   (See p. 108.)

The counties of this group have certain points of

resemblance. All alike extend from the Pennine Moors to the sea-coast, and have therefore the same varied scenery of moorland and plain; all have easy communication with the sea; and all but one share in the industries which gather on the coal-fields of the northern hills.

I. **Northumberland** is the northernmost of English counties; it has an area of nearly 2,000 square miles, and a population of 386,646 persons. In its rude triangular form it is not very unlike a miniature England. The base of this triangle rests on the Tyne and the Derwent to the south; its point to the north is formed by the town of Berwick-on-Twéed; Scotland and Cumberland bound its western side, the North Sea its eastern. The western portion of the shire which lies within these limits is occupied by great mountain ranges, the *Cheviot Hills* on the Scotch border, the *Pennine chain* in that part which adjoins Cumberland; and the county is practically composed of the long sloping moors by which these hills descend to the sea—wild and bleak tracts furrowed by deep river-gorges in whose shelter the little agriculture possible in this soil is carried on. There are only two rivers of any importance—the *Tweed* on its northern border, and the *Tyne* which, with its tributary the *Derwent*, marks its southern boundary.

Northumberland is shut out from many means of wealth and progress by its remote position, its harsh climate, and the barren soil of its mountain sides; and in old times was chiefly important as the border land between England and Scotland. But the discovery of

a great coal-field, which lies happily along the more
level ground on the banks of the Tyne and the adjoining sea-coast, opened out to it a new source of
industry, owing to the ease and cheapness of the trans-
port of its coal to other parts of England by sea.   The
Tyne, for a distance of twelve miles from its mouth,
has been turned into a long succession of towns,
docks, and factories, which have grown up into one
continuous city.   On its northern bank, within the
limits of Northumberland, lie *Newcastle-on-Tyne*, with
130,000 inhabitants, and a trade only inferior to that
of London and Liverpool, *North Shields*, and *Tynemouth;* all engaged principally in the exportation of
coal, but also in the manufacture of the iron, zinc, and
lead found in the mines of the district.   The remaining
towns are, from the character of the county, few and
of little importance. The moorlands are thinly peopled
—here and there a little agricultural town nestles in
a river valley, such as *Hexham* on the upper Tyne,
Morpeth, and *Alnwick;* the rest are mere fishing villages near the coast, where the mouths of streams
form little harbours, such as *Blyth*. The old castle of
Bamborough stands on a sea-cliff to the north, looking
out to *Holy Island* and the little group of the *Farne
Islands.* In the extreme north is the independent town
of *Berwick-on-Tweed*, once a very important Scotch
port, now fallen from its old greatness, but holding a
free position between Scotland and England.

II. **Durham** was originally nothing but a desolate
tract of moorland which served as a *march* or border
between the tribes of Northumberland and those of

Yorkshire.   It is only half the size of Northumberland, having an area of 973 square miles ; like that county its shape is a rough triangle—but a triangle pointing to the west and not the north.   Its broad end rests on the North Sea, while its western extremity just touches Cumberland and Westmoreland. The county consists of three parts—a small portion to the west lies in the moorlands of the Pennine chain and contains lead-mines ; a broad tract of coal-measures stretches from Northumberland across the centre of the shire ; and a belt of undulating agricultural land extends along the coast, in the southern part of which salt-mines are found.

The form of the county is clearly marked by its rivers, which all rise close together in the mountains forming its western point, and all make important harbours at their mouths.   The *Tyne* and its tributary the *Derwent* mark the northern side of the triangle, and the *Tees* the southern, while the *Wear* winds through its centre.   In the valley of this central river, the Wear, lie the old agricultural towns of the county ; *Durham*, the capital, with its old cathedral and castle built on the summit of a mass of rock rising from the river : and to north and south of it the small towns of *Chester-le-Street* and *Bishop-Auckland*.   But all the towns of importance for size and industry lie, as in Northumberland, on the estuaries of the two rivers, the Tyne and the Wear, which pass through the centre of the coal district and carry its trade to the sea. The greatest of these are *Sunderland* with *Monk Wearmouth* and *Bishop Wearmouth*, at the mouth of the Wear, which together have 104,000 inhabitants; *South Shields* on the estuary of the Tyne; and *Gateshead* opposite Newcastle; all of which are maintained by the trade in coal and iron, the manufacture of glass, and ship-building.   On the southern borders of the county lie the towns of the Tees valley: *Barnard Castle* on the upper Tees; nearer the mouth of the river *Darlington* and *Stockton-on-Tees*; and *Hartlepool*, a sea-port to the north of its estuary.   These towns of the south are principally engaged in salt-manufacture, in shipping, and in the making of sail-cloth. Durham has thus many advantages over Northumberland, in the greater variety of its industries, the greater extent of its fertile ground in the river valleys and by the coast, and the number of its sea-ports.   Its population is nearly twice as great as that of Northumberland, being 685,089—though the county is only half as large.

III. **Cumberland** is a county of very irregular form, which lies on the north-western shores of England, and stretches in a slanting direction from the moors of Northumberland on the east to the Irish Sea on the west.   To north and north-west of it are Scotland and the Solway Firth, to the south the district of Furness and Westmoreland.   The county itself is but little smaller than Northumberland, having an area of 1,565 square miles.   It is composed of three distinct parts: the south-western half is very mountainous, containing the greater part of the Cumbrian hills; the north-eastern portion is formed of the wild and desolate slopes of the Pennine moorlands; and between the two lies a broad and fertile plain, the lower valley of the Eden. Cumberland contains the highest mountains in England, and most of the English lakes. The chief mountains are *Scafell Pikes*, and the two great neighbouring heights of *Scafell* and *Bowfell*, lying near its southern borders; and in the centre of the county *Skiddaw* (see page 43). The lakes lie in the hilly country between these mountains. *Derwent Water* and *Bassenthwaite Water* lie under the heights of Skiddaw, *Wast Water* below the Scafell Pikes, and half-way between these are *Buttermere* and *Crummock Water*. The high and beautiful valley of *Borrowdale* stretching from Derwent Water towards Wast Water closes this semicircle of the Cumberland lakes. The only large river of Cumberland is the *Eden*. Next in size is the *Derwent*, which with its tributaries drains four of the lakes. The *Duddon* and many other small mountain torrents are only famous for their wild and beautiful scenery.

As the only industry of the mountainous parts consists of sheep-grazing and slate-quarrying, the population and the large towns were in former times chiefly gathered in the central agricultural plain. Here on the banks of the Eden grew up the chief town, *Carlisle*, with its cathedral, and in later times its cotton manufactures; and some others not a fourth its size, *Penrith* higher up the valley; *Brampton* and *Longtown* east of the Eden; and *Wigton* west of it. Among the mountain villages, *Keswick* on the Derwent alone arrived at any-importance by its lead mines and manufacture of pencils. But in later times the discovery of a bed of coal with iron ore on the sea-coast led to a great increase of population on the western side of the Cumbrian hills, and to the building of new towns entirely maintained by the mining trade, such as *Cockermouth* on the Derwent ; and along the sea-coast, *Maryport*, *Workington*, and *Whitehaven*, the largest of all the towns after Carlisle ; *Egremont*, farther south, lies among the iron mines. The industries of Cumberland, however, are limited to a small part of the county, and its population is very much less than that of its neighbours to the east, being but 220,250.

IV. **Westmoreland**, the smallest of the northern counties, has an area of 758 square miles and is about half the size of Cumberland. It is almost wholly inclosed between that county and Yorkshire ; Durham just touches it on the north-east ; to the south and south-west lie Lancashire and its outlying district, Furness, between which Westmoreland succeeds in penetrating at one point to the sea. The county is very mountainous ; on the east it creeps up the Pennine Moors of Yorkshire ; on its western side it stretches into the heart of the Cumbrian group and includes the second greatest of English mountains, *Helvellyn*, and two large lakes which lie on either side of it, *Ulleswater*, the greatest of all English lakes, on the Cumberland side, and *Windermere* on the Furness side. These heights to east and west of the shire are bound together by a line of lower moorlands running between them, and slightly broken in the centre by the Pass of Shap Fell. To the north of this line the ground becomes comparatively level and forms the upper valley of the river *Eden*. To the south two narrow valleys formed by the rivers *Lune* and *Kent* lie between low ridges of heights, and constitute the square projection which extends from the body of the county southward.

In so hilly a country little agriculture is possible; there is no means of livelihood for the inhabitants save the grazing of mountain sheep and the working of a few mines of copper and lead. Hence Westmoreland with its 65,000 inhabitants is more thinly peopled than any other county in England. It has but two towns in its lower grounds; *Kendal*, which lies in the south on the river Kent near the sea, and has grown to considerable size by manufactures of wool and cotton; and *Appleby* in the Eden valley, not a sixth the size of Kendal but yet holding the place of county town. The village of *Ambleside* at the head of Windermere has become famous to tourists for its beautiful situation.

CHAPTER X.

As we pass from Crossfell southward along the Pennine range, we see two great rivers, the Ribble and the Mersey, thrown down from its western slopes. The basins of these rivers form the bulk of the counties of *Lancashire* and *Cheshire*, which lie along the lowlands of the coast from the Cumbrian to the Welsh mountains, and thus form the western extremity of the great plain which sweeps round the southern part of the Pennine range. (See p. 59.) They therefore share in the mineral wealth of the mountains and the agricultural riches of the plain, while they possess the advantage of easy communication with the sea ; and in industry and population they far surpass the whole group of the northern counties taken together. (See p. 108.)

I. **Lancashire** is a tract of country shut in between the moorlands of Yorkshire on the east and the Irish Sea on the west. Its northern point touches Westmoreland, its broad southern end is bounded by the river Mersey and by Cheshire. In form it is pear-

shaped, narrow at one end and very broad at the other; in size it is about equal to Northumberland, its area being 1,905 square miles. The shire consists of two great divisions of wholly different character and scenery—a mountainous district on the east formed by a succession of moors thrown out from the Pennine chain, and a belt of low and marshy soil running along the sea-coast on the west. The contrast between the two sides may be seen by comparing the highest mountain of the county, *Pendle Hill*, on its eastern border, with the broad, dreary marsh of the *Fylde* to the west.

Lancashire is cleft by three large *rivers;* the *Mersey*, which forms its southern boundary, the *Lune* in the north, and the *Ribble*, which cuts through its centre. The smaller *Wyre* passes through the district of the Fylde, making for itself a large and important opening in the low coast.

The Ribble as it passes through the midst of Lancashire divides it into two parts which are strikingly contrasted in importance, wealth, population, and industry.

(1.) *North Lancashire*, or the district which lies to the north of the Ribble, is agricultural and pastoral, with a very few towns near the sea-coast, and a scanty population. *Lancaster*, the county town, lies at the mouth of the Lune; *Fleetwood*, a sea-port on the Wyre, *Blackpool*, *Lytham*, and *Kirkham*, are all situated near the sea-shore in the midst of a dull, swampy, and unhealthy flat, and are of little importance; there is indeed one great manufacturing town, *Preston* on the

Ribble, but this lies so close to the borders of south Lancashire as to share in its industry and wealth.

(2.) *South Lancashire*, on the other hand, the district between the Ribble and the Mersey, is the richest and most populous part of England, and differs equally from the northern part of the county in the character of its physical structure and in its industrial conditions. It consists mainly of a great three-cornered wedge of moorlands thrown out westward from the Pennine chain, which gradually sink as they near the sea. The broad base of this wedge fills the whole space between the upper valleys of the Ribble and Mersey, while the narrow end extends to a point near St. Helen's.   On the west these moorlands are parted from the sea by a belt of level ground ; on the north they slope rapidly to the narrow valley of the Ribble, and throw down from their steep sides to this river two small streams, the *Calder* and the *Darwen ;* on the south they fall by a gradual descent to the plain watered by the Mersey.   A number of streams which rise in the highest parts of the moorland district are gathered up on its southern slopes by the *Irwell,* and by it carried down to the Mersey.

The tract of moorland which thus forms the framework of south Lancashire is at the same time the main source of its wealth and importance ; for it contains the western half of a vast coal-field which once stretched across England, but has been rent in two by the upheaval of the Pennine range, and now lies half in Lancashire, and half in Yorkshire.   On this great coal-field, which owing to its position on the moorland slopes is traversed by a complete system of running waters, has sprung up one of the chief manufactures of England, that of *cotton*. Before its coal-mines were opened, indeed, the whole district was very rude and wild, and contained but a small pastoral and agricultural population, living in little hamlets hidden in its river valleys. But with the growth of mines and manufactures these hamlets quickly expanded into some of the greatest towns not only of England but of the world; and the whole county has now the appearance of one unbroken city of mills and factories, all busied in the same trade, the weaving, dyeing, and printing of cotton. The towns lie for the most part on the banks of some mountain stream : those situated far up among the moors are comparatively small, while the most important have grown up lower down where the·rivers are larger and their valleys broader. The towns of *Todmorden, Bacup, Haslingden,* and *Wigan,* mark the crest of the heights. On the steep northern slopes of the moorlands, in the basin of the Ribble, are *Colne, Burnley, Accrington, Over Darwen,* and *Chorley.* Nearer to the Ribble lie *Clitheroe, Padiham,* and greater than all these, *Blackburn,* with its 76,000 people, situated in the lower valley of the Darwen. All these towns have a large manufacturing industry. But it is the long southern slope of the coal-fields towards the Mersey that forms the chief seat of the cotton trade, having a more sheltered climate, a more abundant water-supply, and more easy means of river communication with the sea. In these southern valleys lie *Bolton, Bury, Rochdale,*

Heywood, Middleton, Oldham, and *Ashton-under-Lyme*
on the borders of the county, a group of towns which
for the most part contain from 40,000 to 80,000 inhabitants.

All these towns except the last lie near streams
which empty themselves into the Irwell, and which
carry into its broader valley the trade that gathers along
their course.   The banks of the Irwell itself therefore, at a point where three streams meet, have become
the site of a city whose population and wealth exceed
that of the whole group of towns taken together of
which it is the centre.   This is *Manchester,* which with
its suburb, *Salford,* has 480,000 inhabitants, that is,
100,000 more than the whole population of Northumberland.   Lying on the highway which leads from the
moorland valleys to the Mersey, and so to the sea,
and thus gathering into itself the trade of the towns
scattered along these valleys, just as the Irwell gathers
up their streams, Manchester has become the capital
of the trade of south Lancashire, the store-house in
fact of all its manufactured goods, and one of the
most important towns of England.

To the west of Manchester lie other large towns
engaged in the cotton trade, *Leigh, Newton,* and
St. Helens with great manufactures of plate glass ;
and *Warrington* on the Mersey.   Passing yet further
westward along the Mersey, we come to a town even
greater than Manchester, the sea-port of *Liverpool,*
with 500,000 inhabitants, and with a suburb on the
opposite shore of the river, *Birkenhead,* which contains
50,000 people.   The magnificent harbour formed by

the estuary of the Mersey has made Liverpool the outlet for the whole foreign trade of north-western England, and above all for the cotton trade with America; the river is here lined for many miles with docks in which every country of the world is represented, and is only second to the Thames in the extent of its commerce, as Liverpool is only second to London in size and wealth.

The population of Lancashire is 2,819,495; greater than that of Yorkshire, a county three times as large in extent, and much greater than that of the four northern counties together. This is owing to the immense natural advantages which it has above all those shires both for manufactures and for commerce.

To the county of Lancashire belongs also a part of the Cumbrian hill country, called **Furness**, which lies between Westmoreland and Cumberland, and projects into Morecambe Bay. This outlying district has all the characteristics of the Cumbrian group to which it geographically belongs (see p. 43). It includes two of the Cumbrian Lakes—*Coniston Lake* in its centre, and *Windermere* along its borders. The high mountain of *Old Man* rises at the head of Coniston Lake, and from it the hills sink gradually down in long lines to the southern sea-shore, where the island of *Walney* lies in a long low line guarding the coast. The peninsula which stretches south between the river Duddon and Morecambe Bay is very rich in minerals—coal, iron, lead, and copper; and here has grown up a great mining town, *Barrow-in-Furness*, famous for its wonderfully rapid development. In 1846 one hut marked its site, and one-fishing boat lay in its harbour ; in 1874, 40,000 people were gathered together round the rich iron-mines which had been opened in its neighbourhood, and the factories to which they had given rise. *Dalton* and *Ulverston*, a little to the north of it, are but small towns.

II. **Cheshire** is the last of the counties which lie between the Pennine Moors and the Irish Sea, and its character is very different from that of the other shires which we have traversed, breaking away as it does so quickly from the mountainous country and opening into the plain of central England.   It has an area of 1,100 square miles ; its form is that of a casket with two handles, one of which, *Longendale*, runs up into the Pennine moors, while the other, *Wirral*, forms a peninsula reaching out to the Irish Sea.   To the north and east Cheshire is shut in by mountainous country, the moorlands of Lancashire, and the Peak of Derbyshire; to the south it broadens to the plains of Staffordshire and Shropshire ; while the Welsh mountains close it in again on the south-west, so that it has but a narrow opening to the sea between the estuaries of the Dee and the Mersey.

The bulk of Cheshire lies between these rivers, the *Dee* and the *Mersey*, and is cleft in two by the *Weaver*, which flows in a valley occupied by the greatest salt-mines and brine springs of England. The shire consists mainly of a broad plain of red marls, little fitted for tillage, but forming some of the best pasture-land in England, and divided into great dairy and cheese-making farms. A tributary of the Weaver, however, the *Dane*, is thrown down from a tract of rising ground to the east, where the county runs up into the Pennine Moors and has a different character, more like that of the other shires of the northern group. Here are the heights of *Macclesfield Forest* and *Congleton Edge*, outliers of the Pennine chain, with mines of copper and lead, and quarries of building stone ; and here too is a belt of coal-measures which stretches from south Lancashire between the upper valleys of the Mersey and the Dane. It is in this district therefore that the manufactures of Cheshire and the bulk of its population are gathered—in fact there are but two towns of any note to westward of the valley of the Weaver.

The towns of Cheshire group themselves naturally along the river valleys according to their various industries. (1.) Those of the Mersey share in the trade of Lancashire, as *Staleybridge* and *Stockport* with their great cotton factories ; *Altrincham* indeed is a little agricultural town a few miles from the river as it enters on the plain of grazing land; but *Runcorn* at the junction of the Mersey and the Weaver is wholly engaged in the working of iron ; and at the mouth of the river is the port of *Birkenhead*, the busy suburb of Liverpool. (2.) The region on either side of the Dane is the centre of a special trade of Cheshire, the *silk* manu-facture. A few miles to the north of the river is *Macclesfield*, the capital of the silk weaving district ; *Congleton* on the Dane itself, and *Sandbach* to the south of it, carry on the same trade. (3.) The towns

K 2

which lie in the low valley of the Weaver are engaged
in the manufacture of *salt* from the brine springs and
mines of rock-salt for which this valley has been
famous for probably 1800 years.   They are *Nantwich*
on the Weaver, *Middlewich* on the Dane, and *North-
wich* at the meeting of the two rivers.   Near Nantwich is the town of *Crewe*, important only as the
junction of a great network of railways.   (4.) The
last river, the Dee, which skirts the grazing plain to
the south-west, has on its banks the county town *Chester*,
with a cathedral and old castle ; this was an important port till the choking up of the estuary with
sand robbed it of its trade.   The population of
Cheshire is 561,200 persons, and is chiefly gathered in
the manufacturing districts.

CHAPTER XI.

IF we pass southwards over the belt of low undulating country which bounds the plain of Cheshire, we enter directly on the *basin of the Severn* (see p. 59). The shires which occupy this basin we may term from their position the *West Midland Counties*. They form the borderland between England and Wales; indeed of two which lie west of the Severn channel, *Hereford* and *Monmouth*, the last was detached from Wales as late as the reign of Henry VIII. On the eastern side of the river lie the bulk of *Shropshire*, *Worcestershire*, and *Gloucestershire*, while *Warwickshire* occupies the valley of the chief affluent of the Severn, the Avon, and thus forms a narrow wedge which runs up into the heart of England. All these counties lie wholly inland save at the point where the Severn breaks out into the estuary of the Bristol Channel (see p. 108).

I. At the head of the Severn basin, where the river pours down its waters from the Welsh hills and winds round their outliers to the south, lies **Shropshire**—a county with an area of 1,290

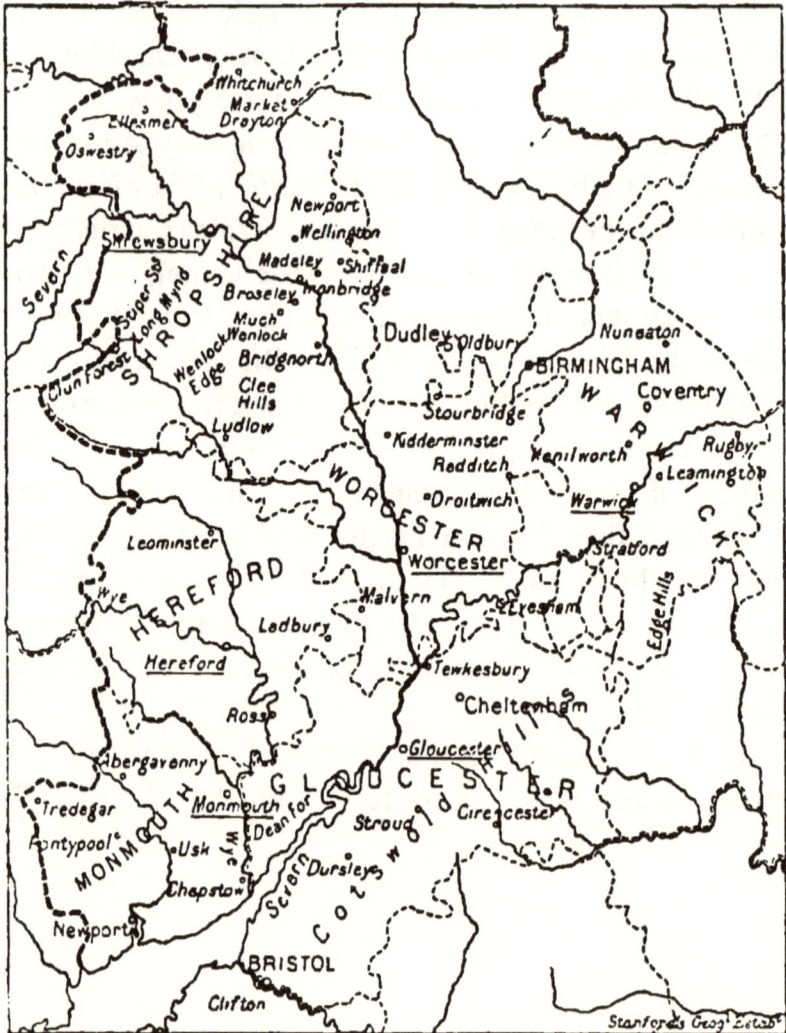

THE COUNTIES OF THE SEVERN BASIN.—Towns and Rivers.

square miles, a little larger than Cheshire.   This shire belongs half to the central plain of England, half to

the Welsh group of mountains. On the north and east
it is bounded by the lowlands of Cheshire and Stafford;
along its western borders rise the hills of Denbigh,
Montgomery, and Radnor, all parts of the Welsh
mountain-group; and on the south is the broken
country of Hereford and Worcester. Shropshire itself
consists of two distinct parts. The northern half,
which adjoins the English plain, is a level tract of rich
pasture-land well watered by rivers; the southern half
is a wild and rugged district of lofty ridges and
hills thrown out from the Welsh mountains.

These two parts are separated by the valley
of the *Severn*, which crosses the county diagonally
and gathers into itself all the rivers of the shire.
These flow to it for the most part through the wide
pasture-fields of the northern plain, as the *Tern*, the
Roden, and the *Perry*. There is but one stream, the
Teme, on the borders of the county, whose course lies
on the southern side of the Severn. As the rivers
with one exception lie to the north of the Severn,
so the mountains with one exception lie to the south
of it. From the heights of *Clun Forest* in the south-
west, long ridges of hill radiate to the north-east like
spokes of a wheel, till cut short by the great Severn
valley—*Shelve Hill, Stiper Stones*, the *Long Mynd*,
the *Caradoc Hills*, and *Wenlock Edge;* to the east of
these the round mass of the *Clee Hills* rises in the
centre of a broad plain, and the high ground of the
Forest of Wyre occupies the south-eastern corner of
the county. The solitary upthrow of the *Wrekin*, 1320
feet high, close to the northern banks of the Severn,

is the one hill which breaks the level plain beyond that river.

The towns of Shropshire are chiefly gathered in the valley of the Severn or its neighbourhood. *Shrewsbury*, the chief town, lies in a loop of the river which almost encircles the ground on which it stands. Further to the east, as its stream passes the Wrekin and winds through the narrow rocky pass of *Coalbrook Dale*, the Severn channel lies across a small belt of coal-measures which forms the manufacturing district of the county, but whose stores are rapidly becoming exhausted. Here *Wellington*, *Madeley*, and *Ironbridge* are all occupied in the iron and coal trade; while *Broseley* is famous for its pipe manufactures. *Bridgnorth*, lower down the river, is like Shrewsbury wholly agricultural. In the great level stretch of rich soil to the north of the Severn valley there are a number of quiet country towns depending on grazing and agricultural industries : *Oswestry, Ellesmere, Whitchurch, Market Drayton*, and *Newport*, all lying near the borders of the county, and *Shiffnal*, whose trade is in corn and cattle for the mining district. Among the bleak hills of the south the population is very scanty; the little town of *Much Wenlock* lies near the Severn, and *Ludlow* shelters itself in the valley of the Teme. The population of the whole county, 248,111 persons, is less than half that of Cheshire with its fruitful plains and many manufactures.

II. If we follow the Severn as it quits Shropshire, flowing mainly to the south-west, we pass into **Wor-**

cestershire, a shire bounded on the north by the plain of Stafford, on the east by that of Warwick, on the north-west and south-west by Shropshire and Hereford, and on the south by Gloucestershire. Worcestershire is a rather small county, having an area of 738 square miles : it coincides for the most part with the deep valley of the middle Severn, as it divides the Welsh mountain group of the west from the plains and uplands of eastern Britain. This valley, about sixteen miles wide, is shut in on the west by the line of the *Abberley* and the *Malvern Hills;* on the east by the *Clent* and the *Lickey Hills*, 1,000 feet and 800 feet in height. Between these two walls of low heights the Severn passes through the county from north to south. On its right it receives the *Teme*, which cuts through the western line of hills to join it ; on its left, as it leaves the southern border, it takes up the much greater river *Avon*, which comes from the heart of England, and which in its passage through the uneven ground of south-eastern Worcestershire, follows a second smaller depression, the *Vale of Evesham.*

The valleys of Worcestershire abound in luxuriant hop-gardens and orchards of apples and pears; but besides its agricultural industries it possesses a variety of manufactures. In the valley of the Severn lie the capital *Worcester*, famous for its porcelain and its gloves ; *Droitwich*, the centre of a salt-mining district which extends along the little river Salwarp ; and *Kidderminster*, grown great by its manufacture of carpets. But the chief manufacturing

towns lie in the north, where the iron and coal trade
of Staffordshire overlaps the border. Here, in a
detached bit òf the county lying in Staffordshire, are
the immense ironworks of *Dudley*, and close by, the
lesser factories of *Oldbury*, and the glass and iron
works of *Stourbridge*. A smaller town, *Redditch*, where
needles and fishhooks are made, is situated at the
foot of the Clent Hills in the east. The quiet little
country town of *Evesham* lies in its vale by the Avon,
and on the western border, at the foot of the
Malvern Hills, is *Malvern*, famous for its mineral
springs. The population of Worcestershire, owing to its
many industries, is large for the size of the county:
it contains 338,837 inhabitants, a number greater than
that of the much bigger county of Shropshire.

III. The Severn crosses the southern border of
Worcestershire near Tewkesbury, and flows in irregular
windings to the head of its estuary through the county
of Gloucester. In this lower part of its course the
Severn valley is bounded on the west by the heights
of *Dean Forest*, and on the east by the long escarpment of the *Cotswold Hills*. Between these heights,
and spreading over them to east and west, lies
Gloucestershire. This shire has an area of 1,258
square miles: to the west it is bounded by Monmouth and Hereford; to the north by Worcestershire; and to the north-west by Warwickshire
—all shires belonging to the basin of the Severn. To
the east lie Oxfordshire and Wiltshire, which belong
to the basin of the Thames, and to the south the
broken country of Somersetshire. Though the *Thames*

rises on the Cotswold slopes within the eastern borders
of the shire, the *Severn* is the only great river which
flows right through the county; in its passage it takes
up on the northern borders the greater *Avon*, and on
the southern boundary the lesser *Avon*.

By its physical structure Gloucestershire is broken
into three well-marked portions. (1.) The first of
these, a tract of red marls bordering the Forest
of Dean on the west of the Severn, is small and
unimportant. (2.) To the east of the Severn, and
between that river and the escarpment of the
Cotswolds, a long narrow strip of country runs
from north to south—a country of tillage and of
orchards for the most part, containing the agricultural town of *Tewkesbury*, and *Gloucester* the chief
town of the county, both on the banks of the Severn;
and on the slopes of the Cotswolds to the east,
the watering-place of *Cheltenham*. The clothing
manufactures of *Stroud* and *Dursley* lie on the range
of the Cotswolds as they extend southwards, overlooking the vale of *Berkeley*. Yet farther south, between
the Cotswolds and the Severn estuary, an outlying
fragment of the Somersetshire coal-field extends across
the southern borders of the shire; and here, on the
little *Avon* as it runs into the Severn estuary, stands the
great town of *Bristol* with its population of 180,000
persons, and its manufactures of glass, sugar, tobacco,
soap, wax, machinery and metal work. In the port of
Bristol nearly all the trade between England and
Ireland is carried on, besides an extensive commerce
with the West Indies, South America, and the Black

Sea. This central strip of low-lying land forms therefore the richest and most populous part of Gloucestershire. (3.) From the Cotswold escarpment eastward
stretches a third tract composed of uplands which
belong to the great oolitic range—a lonely tract of
grazing country, in which *Cirencester* is the only important town. The whole population of Gloucestershire amounts to 534,640 persons, and is chiefly
gathered into its central district by the Severn.

IV. As the oolitic uplands stretch away from the
Cotswolds through Oxfordshire and Northamptonshire,
their escarpment looks down to the north-west over the
valley of the greatest among the eastern affluents of
the Severn. This is the river *Avon*, and the basin
of the Avon forms the bulk of **Warwickshire**, which
thus links itself naturally to the counties of the
Severn.

Warwickshire has an area of 880 square miles.
It is only on its southern and eastern borders that it
belongs to the oolitic range, which skirts it on this side
in Gloucestershire, Oxfordshire, and Northamptonshire;
for the bulk of the county as it stretches on the north-
west to Staffordshire and Leicestershire, and on the
west to Worcestershire, belongs to the plain of central
England. Its scenery answers to this position. On
the southern borders of the county where it adjoins the
upland range there are low hills, as the *Edge Hills ;*
while in all other parts the surface of the ground is only
varied by gentle undulations, which were once covered
by the Forest of Arden, but now form a tract of open
land very fertile and still rich in wood. Across the

centre of this plain lies the depression made by the valley of the *Avon*. Though the *Tame* and a few smaller streams flow northwards to the Trent, it is the *Avon* which gathers in the greater number of the rivers of the shire, and carrying them to the Severn adds Warwickshire to the counties which occupy its basin.

Most of the Warwickshire towns lie in this broad pleasant valley of the Avon. As the river enters the shire it passes near *Rugby*, important for its great school; *Coventry*, a large town supported by silk and ribbon manufactures, lies a few miles to the north of the river; *Kenilworth Castle* is not far from Coventry. At the junction of the *Leam* with the Avon in the centre of the shire is *Leamington*, noted for its mineral springs; and close by, *Warwick*, the chief town of the county; *Stratford-on-Avon*, the birthplace of Shakespeare, lies where the river first touches the southern border of Warwickshire. The towns situated outside the Avon valley are *Nuneaton*, a small place on the northern border, and *Birmingham* in the north-east. This last town is one of the largest in England, containing nearly 350,000 inhabitants; it lies in a little tongue of Warwickshire which runs out into the midst of the coal-fields of south Staffordshire, and shares in the enormous manufacturing prosperity of that region, of which Birmingham has become the trading centre. By its immense factories for steel, iron, needles, pins, plated ware, &c., it has now grown to be more than thirty-three times as great as the capital of the shire. The manufacturing industry which at this point crosses over the border of Warwickshire has raised its population to 634,190 inhabitants, a very large number for the size of the county.

V. The Severn has but one great affluent on the east, whose basin, as we have seen, is occupied by Warwickshire. But on the west its estuary receives two great tributaries, the *Wye* and the *Usk*. The basins of these rivers form the counties of *Hereford* and *Monmouth*.

Herefordshire is a broken, circular plain with an area of 836 square miles, which is shut in on the north by the hills of southern Shropshire, on the west by the high moors and mountains of Radnor and Brecknock, and on the east by the Malvern Hills and the Forest of Dean. The plain thus inclosed on all sides save that towards Monmouthshire on the south, belongs to the basin of the river *Wye*, which flows across its centre, receiving tributaries from the surrounding hills, as the *Lug* from the north and the *Monnow* from the west. The scenery of Herefordshire is varied by gentle hills lying principally along its eastern borders. Its whole industry is agricultural, its abundant streams and great fertility having covered it with hop-gardens and orchards, which have given rise to its main business of cider-making. The population is small, 125,370, and the towns are few. The chief among them are *Hereford* in the centre of the county on the Wye, and north of it *Leominster* on the Lug. *Ross* on the Wye in the south, and *Ledbury* in the east, are both very small country towns.

VI. Through the one open border of Herefordshire to the south, the Wye makes its way in a direct line

to the estuary of the Severn—a line running due north
and south, which forms the eastern boundary of Monmouthshire. This is a very small county, with an area
of only 576 square miles ; it lies between the plain of
Hereford on the north, and the Bristol Channel on the
south, while to the east rise the low heights of Gloucestershire, and on the west the Welsh mountains of
Brecknock and Glamorgan. On the northern and
western borders of the county adjoining the Welsh
mountains are some high outlying hills of the Brecknock range, the highest point of which is the *Sugarloaf
Mount* in the north. These hills fall in long slopes to
the varied plain which forms the greater part of the
shire, a plain which is in fact the basin of the river
Usk. From its source in the Welsh mountains, the
Usk flows across the heart of Monmouthshire on its
way to the estuary of the Severn, thus linking this
shire with the counties of the Severn basin, and forming by its valley the eastern boundary of the coal-field
of South Wales, while it at the same time opens an
outlet to the sea for its mineral wealth.

The towns of the shire lie for the most part along
the river-valleys. The county town *Monmouth* takes
its name from the Monnow, being built at the junction
of that stream with the Wye ; and the little port of
Chepstow lies at the mouth of the Wye. On the
banks of the Usk are situated *Abergavenny, Usk,* and
Newport. This last town is of considerable importance, owing to its being the sea-port of the valley of
the Usk and the lateral valleys to westward of it—a
tract of country which lies within the limits of the

coal-measures of South Wales, and has a large mining and manufacturing industry.   The chief towns of this district are *Tredegar* and *Pontypool*, which are engaged in the working of coal and iron, and whose exports are carried out of the country through Newport.

Owing to its manufacturing industries the little county of Monmouth has a population of 195,500, one larger, that is, than that of Hereford.

CHAPTER XII.

THE basin of the Humber is, as we have seen (p. 59), of very great size, for it extends from Cheshire on the south-west to Durham on the north-east, and thus includes the whole tract of low country which encircles the Pennine range from the Tees to the basins of the Mersey and the Severn. The group of five counties which lies within its limits has therefore a greater area than that of any other group in England —an area of nearly ten thousand square miles. This vast district, however, is divided into two parts which correspond to the two distinct river systems which unite their waters in the Humber, those of the *basin of the Trent*, and the *basin of the Ouse ;* and these parts, with a general resemblance in industrial resources and in population, vary greatly in the size and number of their political divisions, for while the first, with an area of 3,792 square miles is divided into four shires : the second and far larger district, with an area of over 6,000 square miles, forms but a single county.

L

To the east of the counties of the Severn basin
and only parted from them by a low undulating
belt of country, lie the **counties of the basin
of the Trent.** The four shires of this group,
Derbyshire, Staffordshire, Leicestershire and *Nottinghamshire,* form taken together, a rough square
set in the middle of England (see p. 108.) Derbyshire, the centre of the group, is made up of the
heights which lie at the southern extremity of the
Pennine range; Staffordshire, Leicestershire, and
Nottinghamshire form a half-circle of low country
which sweeps round these Derbyshire hills, and opens
on the north-west into the plain of Cheshire, on the
north-east into the plain of York. Hence these
three counties link together the plain of Cheshire
and the plain of York, and themselves belong to
the central plain of England. But they are bound
together in a very marked way by the fact of their
forming the basin of the Trent, for they include
the whole tract of country drained by that river and
by its affluents (see p. 81). Let us suppose a
spectator to be standing on the Peak of Derbyshire
and thence looking out southwards over the plain
which curves round the base of the moorlands, in
which the Trent lies like a great bow. On his right
he will see the river as it rises on the northern
borders of *Staffordshire* and winds through the centre
of that county. Leaving the limits of Staffordshire,
he then sees it bend eastward till its course lies right
in front of him, skirting the borders of *Leicestershire*
and *Derbyshire*, where it takes up rivers sent down to

it from both these counties; and finally he will watch
it turning northwards through *Nottinghamshire* to finish
its course on his left hand after having described a

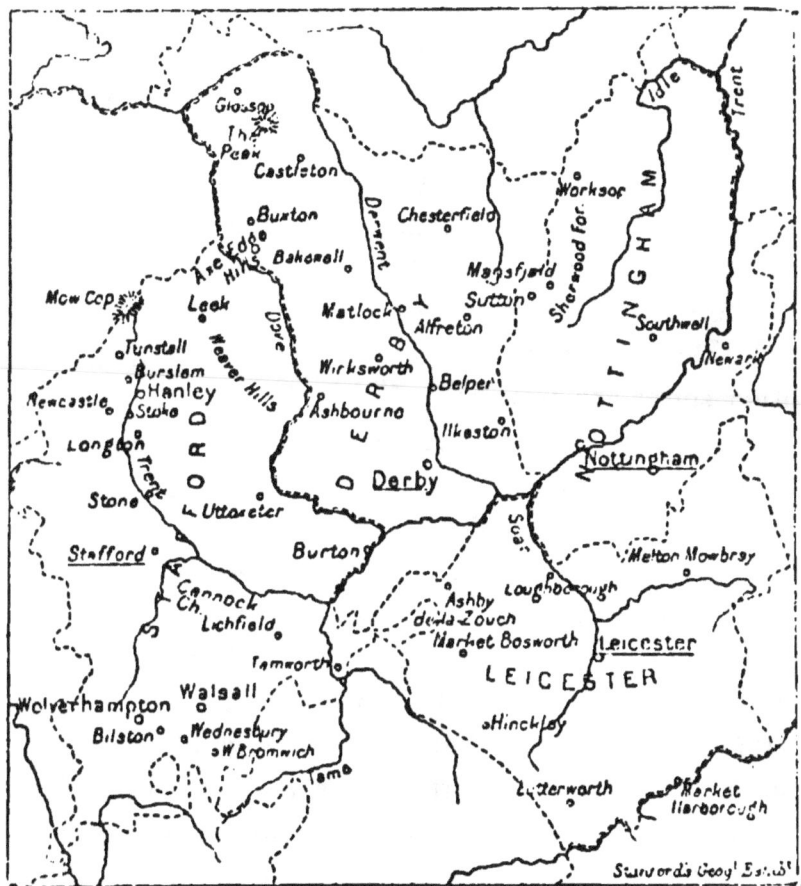

THE COUNTIES OF THE TRENT BASIN.—Towns and Rivers.

complete half circle. The whole basin of the Trent
lies before him, and the four counties of middle
England which occupy this basin are bound together
within it into a natural and sharply defined group.

L 2

The counties of the Trent basin differ from the northern and north-western groups of shires in two ways : (1) they lie wholly inland, and (2) they belong with but one exception to the central plain of England and have therefore little mountainous scenery. In one important point, however, they resemble the northern counties, for encircling so closely as they do the base of the Pennine range, the borders of all four are overlapped by the coal-measures of the moors, and they thus share in the wealth and prosperity of the manufacturing counties of the north and north-west.

I. **Derbyshire** forms the centre round which the other shires are grouped. It is a long narrow county with an area of 1,029 square miles, and a population of 379,400 persons, shut into the midst of England between Nottinghamshire and Leicestershire on one side, and Staffordshire and Cheshire on the other. To the north of it lie the Pennine Moors of York-shire, which are continued through the heart of Derby-shire, and sink gradually towards the valley of the Trent on its southern border. Their greatest heights lie in the northern part of the county, where they form the mountainous district of *the Peak*, a region of rounded hills rising from high moorlands which are cut by wild and deep valleys. But before we reach the southern limits of Derbyshire the Pennine chain, now nearly 200 miles in length, has fallen to the level plain of central England, and here the county extends in a long tongue of low land between Staffordshire and Leicestershire, till it touches the northern borders of Warwickshire. There are three rivers in Derbyshire. The *Trent* just crosses its southern part; but two tributaries of the Trent, thrown down from the Peak, the *Dove* and the *Derwent*, travel for their whole course within the limits of the county. While the *Dove* marks its western boundary, the *Derwent* flows in a deep and beautiful mountain valley through its centre, a valley which forms a natural division breaking the county into two different parts, one manufacturing, the other agricultural.

(*a.*) As the Pennine Moors extend from Yorkshire into Derbyshire they carry with them a continuation of the great coal-measures of Yorkshire, and these coal-measures occupy nearly the whole of that part of Derbyshire which lies to the east of the Derwent valley. Round the coal-mines lie manufacturing towns. *Derby*, the chief town, which is built on the Derwent, near the coal-field, has manufactures of silk and cotton. *Belper*, on the same river, has cotton manufactures. *Chesterfield*, in the centre of the mining country, adds to cotton the Yorkshire manufacture of wool. The working of iron is carried on at *Ilkeston* and *Alfreton* in the east. (*b.*) On the other hand, the hilly country to the west of the Derwent is mainly an agricultural district, where the towns are very small. They are *Ashbourne*, an agricultural town in the valley of the Dove; *Wirksworth*, built near some mines of lead; *Matlock*, on the Derwent, and *Bakewell* and *Buxton* on its tributary the *Wye*, famous for their mineral springs and rocky scenery; and the little village of *Castleton* in the Peak district, where visitors go to see the deep caverns and subterranean streams of the limestone rocks. However, in the north where the county stretches towards Lancashire, it has shared in some of the Lancashire industry, and here the busy town of *Glossop* carries on a large cotton manufacture.

II. Let us now once more take our place on one of the mountains of the Peak, and look out to the south-west towards the source of the Trent. We see that between the plain of Cheshire and the western borders of Derbyshire lies a varied tract of country called **Staffordshire,** which stretches southwards till it touches three counties of the Severn basin, Shropshire, Worcestershire, and Warwickshire. Staffordshire is about the same size as Derbyshire, having an area of 1,138 square miles, and belongs wholly to the central plain of England. Its northern part alone, which lies on the borders of the Peak district of Derbyshire, is broken by some of the last hills and moors thrown out from the Pennine range, such as the heights of the *Axe Edge Hill*, the *Weaver Hills*, and *Mow Cop*. But from this somewhat hilly district to its southern borders the county consists of a gentle undulating tract of agricultural country, with the low upland of *Cannock Chase* in its centre. The whole of this plain is well watered by rivers; the *Trent* passes through the heart of it, receiving on its way many small streams, the chief of which are the *Dove*, which forms the eastern boundary of the county, and on the other side the *Tame*, which crosses it in the south.

. But though the bulk of Staffordshire is from its character agricultural, the great wealth and population of the county.are gathered round two centres near its northern and southern borders. These centres are the two coal-fields of Staffordshire. The northern coal-bed is the end of a long line of coal-measures which run down from Lancashire across Cheshire into north Staffordshire, and which have here become the main seat of the manufacture of earthenware in England, whence the district takes its name of "the *Potteries*." The southern coal-field, separated from the northern by the valley of the Trent, is given up to ironworks and hardware manufactures. These great industries have made Staffordshire one of the richest and most thickly populated counties of England ; it contains 858,326 inhabitants, more than twice as many as Derbyshire, a county of the same size.

The chief town, *Stafford*, lies half way between the coal-fields on the little river Sow; it is principally engaged in the making of boots and shoes. North of it lie "the Potteries" with their great town of *Stoke-on-Trent*, with 130,000 inhabitants, more than nine times as great as Stafford, and its dependent towns *Hanley*, *Burslem*, *Tunstall*, and *Longton*. The neighbouring *Newcastle-under-Lyme* carries on a large trade in hats and shoes. The capital of the iron manufacturing district in the south is *Wolverhampton*, containing 160,000 inhabitants, the largest town of the county. *Walsall* is but little smaller, and after it come *Bilston*, *Wednesbury*, and *West Bromwich*. *Burton-on-Trent* on the eastern borders has the greatest

breweries in England. Then follow some quiet agricultural towns, *Lichfield* with its old cathedral, and
Tamworth in the south; in the north the large town
of *Leek*, and below it *Uttoxeter*, and *Stone* on the
Trent.

III. Looking directly south from our old point of
view on the Peak, we see Derbyshire pushing out a
thin tongue of land which parts Staffordshire on the
south-west from the county of **Leicestershire** on
the south-east, a county which lies on the farther
side of the Trent, whose course here runs from west
to east and just touches the northern border of the
shire. In extent Leicestershire is the smallest county
of the Midland group, having an area of but 800
square miles; its form is not unlike that of a heart.
On the east and south-east it is shut in by the uplands
of Lincolnshire, Northamptonshire, and Rutland; the
southern strip of Derbyshire bounds it on the west;
to the north it opens on the lowlands of Nottinghamshire; to the south on those of Warwickshire.

The shire lies wholly within the plain of central
England, and is chiefly composed of the valley of
the river *Soar*, which rises near its southern point,
and traverses its centre to join the *Trent* on its
northern borders. Thus the middle part of the county
is very low, while on either side the ground slightly
rises: on the east it creeps up the slopes of the oolitic
uplands, which as we have seen (see p. 59) form
the boundary between the Trent basin and that of the
Wash; and on the west it rises to the hilly and somewhat rugged tract of *Charnwood Forest*, and a small

bed of coal which lies beyond it. The streams from the rising grounds to east and west are thrown down to the Soar in the middle, and by it carried to the Trent. The two largest towns lie in the valley of the Soar—*Leicester*, with very great manufactures of woollen stockings and gloves and of woollen and cotton thread; and farther north, *Loughborough*, also with manufactures of hosiery and lace. On the rising ground to westward of the Soar is *Ashby-de-la-Zouch*, a small town in the centre of the coal district. On the uplands of the east just facing it is *Melton Mowbray*, an important agricultural town. A few quiet little country towns are dotted over the plain in the south, *Market Harborough, Lutterworth, Hinckley,* and *Market Bosworth*. The county contains 269,300 inhabitants.

IV. Right to the east as we look from the Peak lies **Nottinghamshire**, a long narrow belt of low undulating country, pent in between the uplands of Lincolnshire on the east and the broken heights of Derbyshire on the west, while to the north it opens into the plain of York, and to the south into the more broken plain of Leicestershire. Nottinghamshire is scarcely larger than Leicestershire, having an area of 822 square miles. It consists of two clearly-marked districts. (*a.*) To the west is a country broken by heights pushed out from the Derbyshire hills, amongst which lies the part once known as *Sherwood Forest;* this district is overlapped by outliers of the Derbyshire coal-measures, which here give rise to some manufacturing industry. (*b.*) The eastern part of the county, on the other hand, is an agricultural plain watered by the *Trent* and by its affluents, thrown down for the most part from the rising ground on the west. The chief of these is the *Idle*, which in its winding course gathers up nearly all the streams of the west, and carries them to the Trent, on the northern border of the county.

The towns of Nottinghamshire are few and chiefly agricultural. The capital indeed, *Nottingham*, which lies on the borders of the coal-field, is a manufacturing town of great size, where shoes, cotton, stockings, and lace are chiefly made; and one other manufacturing town, *Sutton-in-Ashfield*, lies just within the western limits of the shire. *Mansfield*, *Worksop*, and *Southwell* are smaller towns lying in the plain to westward of the Trent; but the chief agricultural centre is *Newark-on-Trent*, where the great corn and cattle markets of the county are held. The population of the county is large for its size, amounting to 319,758 inhabitants.

V. The second great affluent of the Humber is the Ouse, and the **basin of the Ouse** forms **Yorkshire**, the largest of all the English counties, having an area of above 6,000 square miles, and thus covering more space than all the shires of the northern group taken together. To the north of it lies Durham, to the west are Westmoreland and Lancashire, to the east the German Ocean; while three counties of central England, Lincoln, Nottingham, and Derby, lie along its southern borders (see p. 108).

Right across the great tract which lies within these

limits a broad plain stretches from north to south, to which the city of York in its centre gives the name of the *plain of York.* To right and left of this plain lie two much smaller valleys which start from its centre and run to eastward and westward—the *Vale of Pick-*

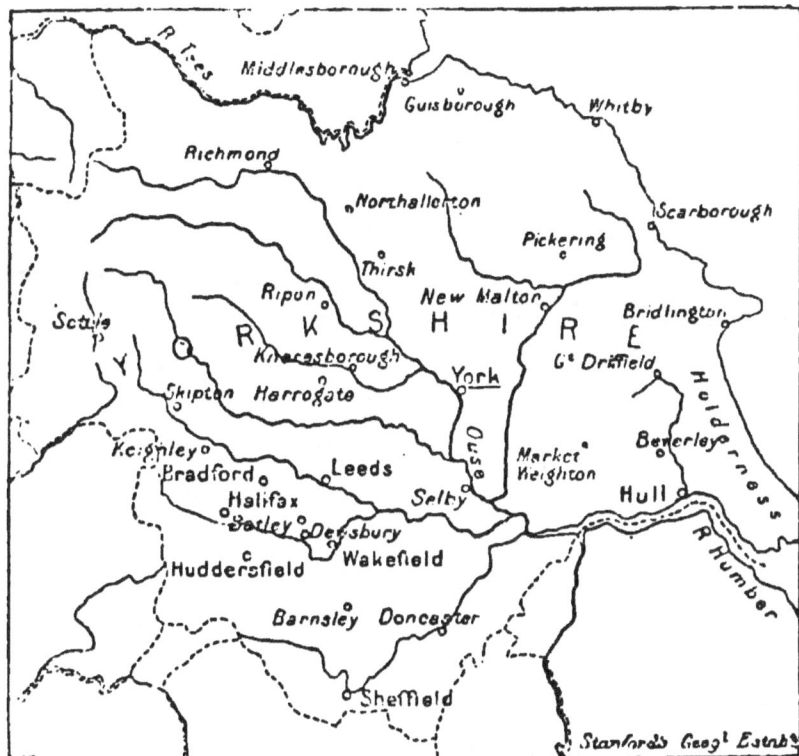

YORKSHIRE.—Towns and Rivers.

ering on the east, and the valley of *Settle* in the west. These three plains lie like a gigantic cross on the face of Yorkshire, and break up the wild hills and moorlands which form its eastern and western portions into four distinct masses. To the north-east lie the

Yorkshire and Cleveland Moors ; to the south-east, parted from them by the Vale of Pickering, are the *Yorkshire Wolds ;* across the plain of York to the north-west are the *Richmond and Craven Moors ;* and south of them, beyond the valley of Settle, the long reach of the *Pennine moorlands* stretching as far as Sheffield.

It is in the north-western moorlands of Richmond and Craven that the chief rivers of Yorkshire take their rise. Thence come the *Ure,* the *Swale,* the *Nidd,* the *Wharfe,* and the *Aire,* which unite their waters as they fall into the plain of York. The *Don* is the only stream of any size which rises in the moorlands of the south-west ; and the *Derwent* alone is thrown down from the York moors of the north-east. One special characteristic of Yorkshire is the way in which all these streams are gathered up into a single channel, that of the *Ouse,* and poured out into the sea by a single estuary, the *Humber.* Thus nearly the whole county falls within the basin of the Humber, for it touches but two rivers of any size not united to this river group—the *Tees* which forms its northern border, and the *Ribble,* whose head-waters lie among the western mountains, but whose course soon turns away out of Yorkshire into Lancashire.

Each of the districts into which we have divided Yorkshire has its own special character. Let us look (1) at the plains and (2) at the mountainous regions.

1. (*a.*) The **Plain of York** is formed of the richest soil in the county; it is traversed by the Ouse, the main water-channel into which the rivers from the

moors are poured; and it has always been the chief
line of communication between England and Scotland.
In the centre of this plain therefore, by the banks of
the Ouse where it first becomes navigable, sprang up
the chief town of Yorkshire, *York*, a city which com-
manded at once the fertile plain around it, the commerce of the Ouse, and the road to Scotland, and thus
became the capital of northern England. The plain
to the north of York is dotted by small agricultural
towns; *Thirsk* and *Northallerton* to eastward on the
Swale; *Ripon*, on the banks of the Ure, the seat of a
bishopric but otherwise of little note; in the valley of
the Nidd *Knaresborough*, containing one of the chief
corn-markets of the north, and *Harrogate*, an important watering-place. To the south of York, near
the borders of the county, is a much greater agricultural town than these, *Doncaster*, situated on the Don,
in one of the most fertile districts of England. The
towns which lie along the banks of the Ouse below
York are engaged in the river trade, as *Selby* with
its shipyards, and in a far greater degree, *Hull* on
the Humber, the capital of the northern trade with
Europe. The plain of York therefore, with its river
opening to the sea, represents the chief agricultural
and trading industries of north-eastern England.

(*b.*) The **Vale of Pickering** is a fertile and well-
cultivated valley shut in between the York moors and
wolds, and watered by the river Derwent. Through it
are scattered quiet country towns and villages such as
New Malton, which marks the old ford across the
river, and *Pickering*.

(*c.*) The **Valley of Settle**, and the adjoining table-lands lying in the heart of the Craven moors comprise *Ribbledale, Wharfedale,* and *Airedale,* or the upper valleys of these three rivers. The valley of the Aire is one of the most fruitful tracts of Yorkshire; the hill-sides which bound it form great sheep-farms, and *Skipton* on the river-banks, the chief town of this district, is noted for its cattle-markets.

2. The mountainous districts of Yorkshire have their characteristic features very distinctly marked.

(*a.*) The **York Moors** are a line of wild and barren uplands running east and west from the plain of York to the sea (p. 59). The *Hambleton Hills* form their western escarpment, overlooking the plain of York, out of which they rise like great inland cliffs. Their northern face is formed by the *Cleveland Hills,* where they slope to the river Tees. Since the discovery of iron-mines in this district about thirty years ago, a large mining population has gathered on the hill-sides. The port of *Middlesborough* on the Tees, and *Guisborough* at the foot of the hills, are the chief centres of their trade, the first being engaged in the export, the second in the mining trade. But the greater part of the York moors is a wild and lonely country wholly given up to sheep-grazing, and contains only a few hamlets in the narrow river valleys which run down to the Vale of Pickering, and small sea-ports or watering-places on the cliffs of the coast, such as *Whitby* and *Scarborough.*

(*b.*) The **York Wolds** form a great crescent of chalk hills sweeping round from Flamborough Head to the

Humber, and bounded on the east by the low ground of *Holderness Marsh*, on the north by the Vale of Pickering, and on the west by the Plain of York. To the north and west the hills rise, like the Hambleton Hills, in steep smooth cliffs lifted out of the plain as out of a great sea, and curved into promontories and bays. The Wold towns are very few and wholly engaged in agriculture ; all except *Market Weighton* lie at the foot of the uplands on the borders of the low and more fertile ground. The chief among them, *Bridlington*, *Great Driffield*, *Beverley*, and the sea-port of *Hull*, mark the inner side of the crescent of the Wolds.

(*c.*) The **Richmond and Craven Moors** of north-western Yorkshire consist of desolate tracts of fern and heather broken by crags of blackened stone, and by the steepest and wildest of the Pennine mountains, *Mickle Fell*, *Ingleborough*, *Pen-y-ghent*, and *Whernside*, among whose heights all the chief rivers of Yorkshire take their rise. The little town of *Richmond* finds shelter at the opening of *Swaledale*.

(*d.*) The **South-western Moorlands**, or that part of the Pennine chain which lies between Settle and Sheffield, form a most important district. Here along the moorland slopes is spread a vast coal-field, where the chief population, wealth, and trade of Yorkshire have been gathered. The abundant supply of coal and iron, and of running waters by which the machinery of mills and factories may be worked, has given rise to some of the most flourishing manufactures of England, those of wool, iron, steel, and cutlery. The towns busied in these trades are thickly crowded together so that the whole district has the appearance of one great city. The chief among them is *Leeds* on the Aire, a town with 260,000 inhabitants, six times as large as York; close to it is *Bradford*, the next in size, with 150,000 people; a little to the north of these lies *Keighley*, and to the south a whole group, *Halifax*, *Huddersfield*, *Batley*, *Dewsbury*, and *Wakefield*. These large towns have populations varying from 15,000 to 70,000; and with a number of lesser ones are all engaged in the manufacture of wool, which is here carried on for the whole of the inland and foreign trade of England. Two other manufacturing towns lie yet further south having different industries; *Barnsley*, carrying on a trade in linen; and *Sheffield* on the Don, at a point where five streams meet, almost as great a city as Leeds, with a population of 240,000 people, and possessing the greatest cutlery manufactures in the world.

Political Divisions.—Yorkshire is divided for political purposes into three great divisions or *Ridings*, all of which meet at the chief town, York. The *North Riding* stretches across the whole upper part of the county, dipping down southward at its broadest part as far as the city of York in the plain of the Ouse. It contains the Cleveland Hills and York Moors, a small part of the south-western Moors, and the plain lying between them. The *East Riding* lies between the Ouse, the Derwent, and the sea and contains the York Wolds. The *West Riding*, the largest of the three, comprises the high moorlands of the Pennine range and that part of the York plain which lies to the west of the Ouse.

As Yorkshire is the largest of English counties, so it is one of the most thickly peopled, its population being nearly 2,436,000. There is but one county in the north, Lancashire, which has a greater number of inhabitants.

CHAPTER XIII.

THE COUNTIES OF THE WASH, OR EAST MIDLAND COUNTIES.

EASTWARD of the counties of the Trent basin lies the group of the *Counties of the Wash*, or the *East Midland Counties*. We have already sketched that great inlet into the eastern coast which is called the Wash, the common estuary of four rivers, the Witham, the Welland, the Nen, and the Ouse (see p. 81). If we take a point on the sea-shore at the mouth of the Nen, we may gain a general view of the *basin of the Wash* (see p. 59). Before us stretches a tract of alluvial soil, whose greatest length, from Lincoln to Cambridge, is seventy-three miles, and whose greatest breadth, from Peterborough to the mouth of the Ouse, is thirty-six miles. This wide tract, the largest level plain in England, has an area of 1,300 square miles, and forms what is known as the *Fenland*, a district of low marshy soil traversed by long canals, or by rivers well-nigh as sluggish as canals, whose courses are marked by lines of willows as far as the eye can reach over the flats. Beyond the fens to the west rise the gentle heights of the Oolitic uplands, which part the basin of the Wash from those of the Trent and the Severn; while to the south lies the steep escarpment of the East Anglian heights which sever it from the valley of the Thames.

The six counties inclosed in the basin between the crest of these uplands and the sea, and which drain their waters into the Wash by means of its four great rivers, radiate outwards like the spokes of a wheel in a vast half-circle, the narrow end of each county being turned towards the Wash—*Lincoln-shire* to the north, *Rutland* to the west, *North-amptonshire* to the south-west, *Bedfordshire* and *Hunt-ingdonshire* inclined yet more towards the south, and *Cambridgeshire* due south (see p. 108). Lying wholly apart as they do from the manufacturing districts of England, and depending solely upon agriculture, their scenery and industry are very different from those of the shires we have already considered.

I. If we now look northwards from our station at the mouth of the Nen we see stretching from the spot where we stand to the Humber the county of Lincolnshire, a great tract of land with an area of 2,776 square miles, about half the size of Yorkshire, and second only to it among English counties. To the north it is severed by the Humber from York-shire; to the west it is bounded by the counties of Nottingham and Leicester; to the east by the German Ocean, and to the south it reaches to Cam-bridgeshire, Northamptonshire, and Rutland. The shire which lies within these limits is made up of two

parts of different character and scenery, (1) the *Up-lands*, and (2) the *Fens*. (1.) The Uplands are divided into two distinct branches which unite on the shares of the Humber. The Lincoln uplands, or the *High Dyke*, form a low rise of oolitic rocks, which run along the western border of the county, connecting the York Wolds with the Northampton Heights; while a second branch, called the *Lincoln Wolds*, consists of a tract of chalk downs which stretch from the Wash in a north-westerly direction to the Humber. By these uplands Lincolnshire forms part of the oolitic range. (2.) The low Fen country, a large part of which is known as *Holland*, or the hollow land, lines the whole of the sea-shore from the Wash to the Humber, and extends between the Wolds and the High Dyke northwards. By this tract of marshy land Lincolnshire belongs to the plain which separates the oolitic from the chalk ranges (see p. 59).

The rivers of Lincolnshire are the *Welland*, which comes from Rutland to wind through the marshes of the south, and the *Witham*, which flows through the centre of the county between its two branches of up-lands, and enters the Wash close by the Welland. Two other rivers which lie outside the basin of the Wash skirt the northern and north-eastern borders of the county, the *Humber*, and the *Trent*.

The industry of the shire is wholly agricultural: its uplands are given up to sheep-grazing, and its marshes have been drained for tillage by immense and long-continued labour. It contains 436,600 inhabitants about one-sixth of the population of Yorkshire, and has few towns of any size. These are mainly agricultural, and lie along the river-valleys. In the fen country by the Welland are the little agricultural towns of *Stamford* and *Spalding*. On the upper Witham lies *Grantham;* the capital and cathedral city of the county, *Lincoln*, with 26,700 inhabitants, lies on the slope of a gap cut by the Witham through the oolitic rocks of the High Dyke; while *Boston*, at its mouth, carries on a shipping trade with northern Europe. *Horncastle* is a small town situated at the foot of the Wolds to eastward of the Witham, and *Louth*, famous for its carpet manufactures, lies on the chalk uplands. The remaining towns are situated on the borders of the county in the basin of the Humber; they are *Gainsborough* on the Trent ; *Glanford Brigg* on an affluent of the Humber; and on the estuary itself *Barton*, and *Grimsby*, a trading port.

II. If we now look due west from our station on the Wash, we see a very small three-cornered tract of land, tightly wedged in between Lincolnshire and Northamptonshire, and bounded by Leicestershire on the west. This is **Rutland,** the smallest of English counties, with an area of 150 square miles. Lying on the oolitic range of uplands, it forms a fertile tract of undulating land watered by the *Welland*, which marks its southern borders, and by some small tributary streams. Its population is small, 22,073, and its quiet agricultural towns are little more than villages,— the chief among them are *Oakham*, in the *Vale of Catmoss*, and *Uppingham*.

III. Looking from the Wash straight to the south-

west the eye runs along a narrow mass of uplands
which is in fact the central portion of the oolitic range.
This long narrow mass is **Northamptonshire**,
whose area is 985 square miles, not very much more
than a third of Lincolnshire. It lies isolated among
lowlands; to the north-west it is bounded by the
plains of Warwickshire and Leicestershire, and to the
south-east by the low broken country of Buckinghamshire, Bedfordshire, Huntingdonshire, and Cambridgeshire. Only at the two narrow extremities of the
country the oolitic uplands of which it is composed
are prolonged northwards through Lincolnshire, and
to the south-west through Oxfordshire.

N orthamptonshire thus consists of a tract of rising
ground bounded on either side by low plains; and by
this clearly marked structure of the county the direction and character of its river-system is determined.
The uplands of the shire form a part of the water
parting of central England (see p. 21) where many
of its greatest rivers take their rise. Hence all the
rivers which traverse the county flow from the midst
of it *outwards*, while not one is able to pass *into* it
from the surrounding plains. There are two chief
groups of headwaters in the shire. The first lies in
the north-west, where the springs of three large rivers,
the *Welland*, the *Nen*, and the *Avon*, rise close
together, the Avon falling down westward to the
Severn, the Welland and the Nen eastward to the
Wash, along opposite borders of the county. The
second group lies in the southern part of the shire
near Daventry, where the *Ouse* and the *Cherwell* take

their rise, the Ouse flowing to the Wash, the Cherwell to the Thames. As the rivers which fall into the Wash flow through Northamptonshire almost from end to end, while little more than the springs of the Avon and Cherwell lie within its limits, the county does in fact belong to the basin of the Wash.

The uplands of the shire are mainly given up to cattle-grazing. In a few places the working of iron-stone found in the soil is carried on, but the chief manufacture is that of boots and shoes. The centre of this trade is *Northampton*, a large town on the banks of the Nen. All the principal towns lie in the basin of the same river—*Daventry* in the west, *Wel-lingborough* not far below Northampton, and *Kettering* a few miles further north, on a tributary of the Nen. A more ancient and famous town than these is *Peterborough*, which lies on the Nen just as it forsakes the county; it is noted for its great cathedral and gives its name to a bishopric. The population of Northamptonshire is 243,890.

IV. If we look from the Wash yet more to the south-west, we see a belt of low clay ground running below the oolitic range of Northamptonshire. This is the county of **Huntingdon**, one of the smallest in England, with an area of only 361 square miles. The low clay soil of which it is composed stretches beyond it in a south-westerly direction into Bedford-shire, and widens into Cambridgeshire in the east.

Huntingdonshire is for the most part a level and monotonous agricultural plain that stretches from the *Nen* in the north to the *Ouse* in the south, both of which rivers belong to the basin of the Wash. The population of the county, 63,700, is smaller than that of any shire in England save Rutland, and its towns are little more than agricultural villages. The most important of these lie in the valley of the Ouse—*S. Neot's* where the river enters the county, *Huntingdon*, the county town, in the centre of its course, and *S. Ives* where it leaves the shire.

V. If we still look in the same direction from the Wash, and if our sight could reach across Huntingdonshire, we should see beyond it a tract of low country with the oolitic uplands of Northampton still lying to the north-west of it, while to the south-east rises the steep chalk escarpment of the Chiltern Hills as they traverse the counties of Hertford and Cambridge. The depression between these ranges is occupied by a fertile plain of 462 square miles in area, which forms the county of **Bedford**. On the south-west the low ground is prolonged into Buckinghamshire, and on the north-west into Huntingdonshire; in fact the depression in which these three counties lie forms but one valley, the basin of the *Great Ouse* (see p. 59).

This river winds in a very irregular course through the centre of Bedfordshire, gathering up as it goes a number of streams thrown down from the Chiltern Hills. The fertile soil thus richly watered has become a well-cultivated agricultural district; the only manufacture is that of straw-plaiting, a trade which gives employment to a considerable number of people and is carried on in most of the towns. The chief town of

the county is *Bedford* on the Ouse in its centre ; *Luton,*
Dunstable, and *Leighton Buzzard* lie on its southern
borders, and the little agricultural town of *Biggles-*
wade in the east.   The whole population amounts to
146,260, which is more than twice that of the shire of
Huntingdon.

VI. As we look from the Wash directly to the
south, we see at the mouth of the belt of low country
which we have just described the shire of **Cam-**
bridge.   Here the valley of the Ouse opens
out into the Fenland which surrounds the Wash, and
the county of Cambridge lies for the most part within
that district.   It has an area of 818 square miles : its
form is roughly speaking that of a crescent which
half encircles the little shire of Huntingdon on its
western side.   In the north it runs up into the Fens
of Lincolnshire ; its southern side is bounded by the
chalk uplands of Hertfordshire and Essex, and its
eastern by the uplands of Norfolk and Suffolk.   The
two great rivers of the county, the *Nen* on the north,
and the *Great Ouse* which crosses the centre of the
shire, come to it from Huntingdonshire.   The basins
of these rivers form two distinct districts.

(*a*) The district round the Nen consists of a low
unhealthy marsh which forms part of the Fen country,
and has only been made fit for habitation and cultivation by laborious drainage.   A part of it, which lies
between the Nen and the Ouse, the *Isle of Ely,* was
formerly a knoll surrounded by low swamps wholly
uninhabited, in the midst of which rose the very
ancient town and cathedral of *Ely* on the banks of

the Ouse. The whole of this district is even now very thinly peopled, containing only the small town of *March*, besides *Wisbeach* on the northern borders of the county, which is important for its corn and cattle markets.

(*b*) The Ouse drains the more populous and healthy plain which lies to the south—a plain given up to grazing and dairy farms, and thickly studded with small villages. The chief tributary of the Ouse is the *Cam*, which gives its name to the county town of *Cambridge* on its banks, the seat of one of the two great English Universities.

The small town of *Newmarket*, chiefly known for its horse-races, lies on the eastern border of the county. In consequence of its disadvantages of soil and climate, the population of Cambridgeshire, 186,900, is very small for the size of the county; it has scarcely more inhabitants than Bedfordshire, whose area is but a little greater than half that of Cambridgeshire.

CHAPTER XIV.

 THE two Eastern Counties, *Norfolk* and *Suffolk*, lie
along the shores of the German Ocean between the
Wash and the estuary of the Thames, and derive their
names from the tribe of the *East Anglian* people who
first conquered and settled this district (p. 108). They
belong so far as the bulk of the country is concerned
to that range of chalk which traverses England from
the Dorset Heights to Hunstanton Cliff on the shores
of the Wash ; and which in its passage through these
counties takes the name of the *East Anglian Heights*
(p. 21). But while these chalk uplands occupy the
western part of Norfolk and Suffolk, their eastern
part is formed of the low reaches of sand, gravel, and
clay which here line the shore of the North Sea.

From the similarity between the Eastern Counties
in position, in structure, and in the character of their
soil, it follows that they are also alike in their history,
their climate, and their industrial pursuits. They were
peopled by the same race, they share in the same dry
and cold climate, are open to the same sharp easterly

winds, and have the same agricultural industries, the same shipping-trade in the towns of the coast, and the same small manufacture of silk-crape and baize

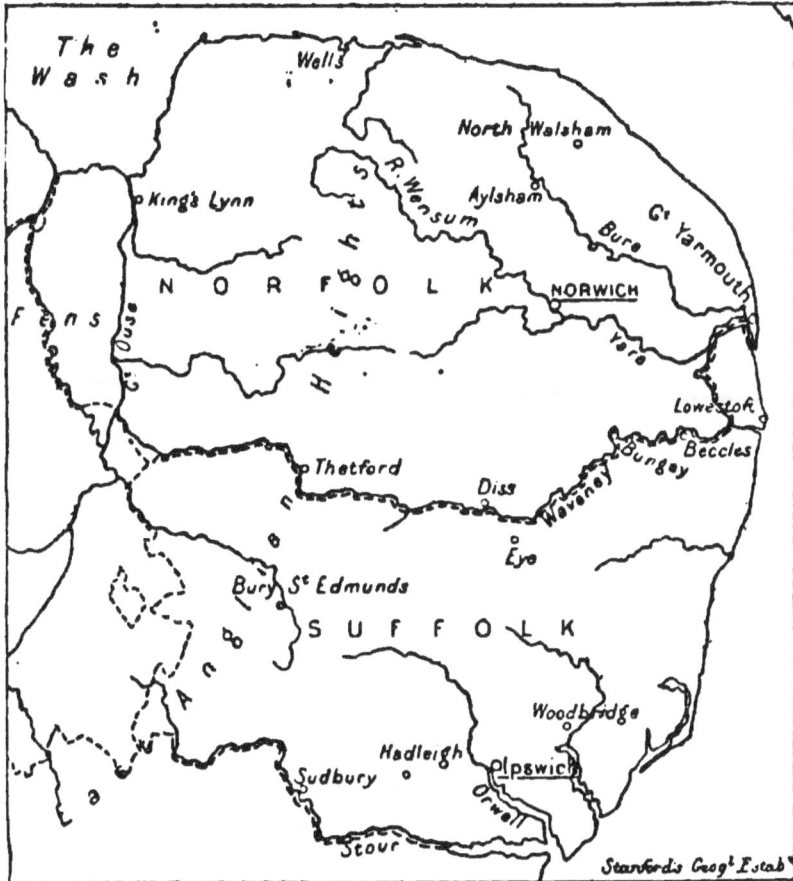

EAST ANGLIAN COUNTIES.—Towns and Rivers.

which they received from the neighbouring European countries. All their large towns lie by the sea-side, where they can carry on fisheries and a little coasting trade, which is limited by the fact that the harbours along their shores are too small to admit vessels of any considerable size.

I. **Norfolk**, the home of the "north folk," or that tribe of the Anglian invaders whose settlements lay to the North, is one of the largest counties in England, with an area of 2,116 square miles.   It is bounded on the north and east by the North Sea; on the south by the river Waveney which parts it from Suffolk; and on the west by the lowlands of Cambridgeshire.   The greater part of the county is occupied by the East Anglian Heights, whose steep western escarpment overlooks a strip of the level Fen country which is included within the limits of Norfolk, and through which the *Great Ouse* winds sluggishly to empty itself into the Wash by the port of *King's Lynn*.   All the other rivers of Norfolk rise among the uplands of the county itself, and are thence thrown down their gentle slopes eastward to the North Sea, traversing on their way a belt of alluvial soil which stretches along the sea and forms the eastern portion of the shire.   The chief of these, the *Bure*, the *Yare*, and the *Waveney*, have the same opening into the sea at *Yarmouth*, the great sea-port of the county.

Norfolk is not very thickly peopled; its population of 438,656 inhabitants is about the same as that of Lincolnshire.   The chief town, *Norwich*, which lies on the Wensum, a tributary of the Yare, is a cathedral city, with 80,000 inhabitants, and large manufactures of silk and wool: *Yarmouth*, on the Yare, its trading port, contains more than 40,000 people. But its other towns, *Wells* on the north coast,

North Walsham and *Aylsham* in the valley of the Bure, and *Thetford* and *Diss* on the southern borders of the county, are little more than agricultural villages.

II. **Suffolk**, the kingdom of the "south folk," lies between Norfolk and Essex, from which counties it is severed by the rivers Waveney and Stour ; and is bounded on the west by Cambridgeshire, and on the east by the North Sea. It is not so large as Norfolk, having an area of 1,480 square miles ; and like that county is partly occupied by the broad undulating downs of the East Anglian Heights, and partly by the belt of low alluvial land which borders the sea. Its rivers, which rise in the upland district, flow directly to the North Sea—the *Waveney* and the *Stour* on the northern and southern borders of the county, and the *Orwell*, which unites with the Stour at its mouth to form a harbour at the southern boundary of Suffolk, answering to that at the mouth of the Waveney at its northern extremity. These harbours, however, lie outside the limits of Suffolk, and the trade of this shire is confined to one seaport town, *Lowestoft*, with very large herring fisheries. The chief industry of the county is agricultural : its most important town, *Ipswich* on the Orwell, has 43,000 inhabitants, and large manufactures for the machinery needed in tillage ; the agricultural towns are *Bury S. Edmunds* on the chalk downs of the west ; *Eye*, *Bungay*, and *Beccles*, in the valley of the Waveney ; and two others near the Stour in the south, *Sudbury* and *Hadleigh*. Though Suffolk is little more than half the size of Norfolk, and contains no single town so large as Norwich, yet its whole population, which is more evenly distributed over the surface of the county, amounts to 348,870, a number nearly as great as that of Norfolk.

CHAPTER XV.

To the south of the counties of the West Midland group lies a line of shires stretching eastward and westward along the *basin of the Thames* (see p. 108). We have already described the remarkable structure of this basin in its upper course between the oolitic and chalk ranges, and in that broader and more important part of its valley which is inclosed between the range of the Chilterns and East Anglian Heights, and that of the Hampshire Downs and Wealden Heights (see p. 59). From the point where the river quits Gloucestershire, it forms a continuous boundary between shire and shire; and, unlike the Trent or the Severn, it in no case enters within the limits of any one of them. On its northern banks are ranged the counties of *Essex*, *Middlesex*, and *Hertford*, which stretch from the river up to the East Anglian and Chiltern Hills that bound its valley on the north, and *Buckinghamshire* and *Oxfordshire*, which extend beyond the chalk range into the upper valley of the Thames among the oolitic clays. On its southern

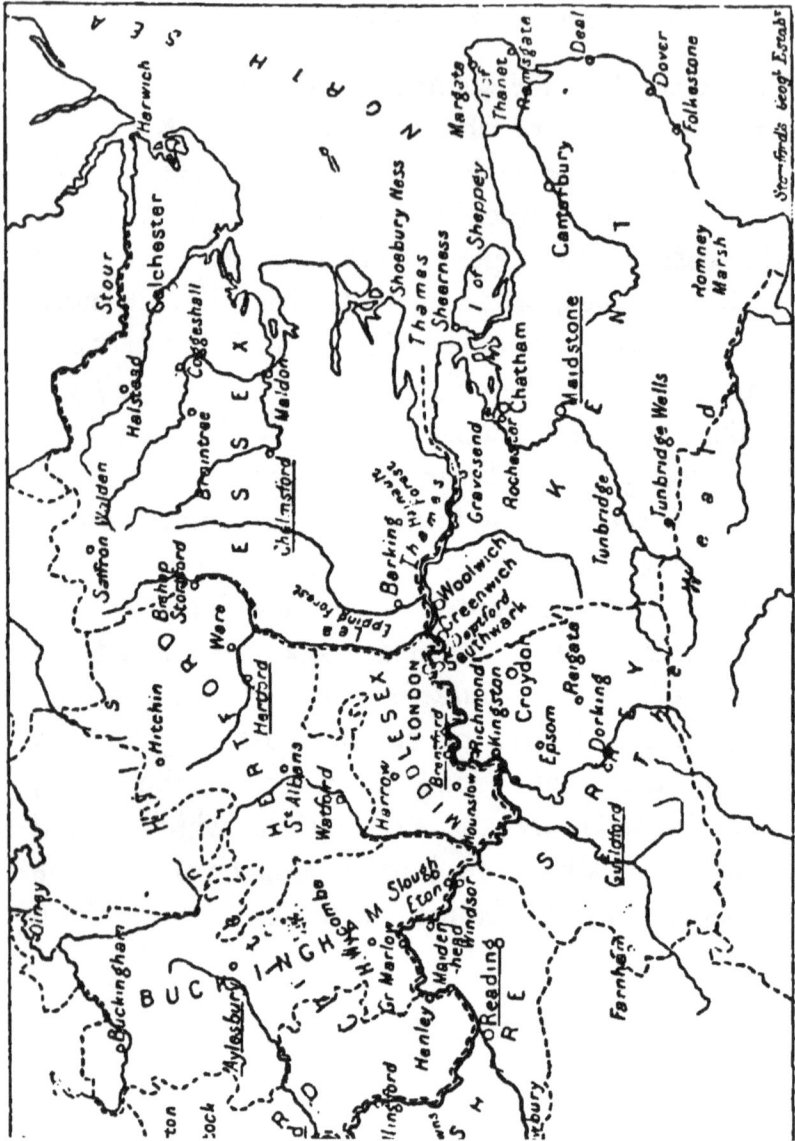

THE COUNTIES OF THE THAMES BASIN.—Towns and Rivers,

banks, in reverse order, lie *Berkshire, Surrey,* and *Kent,* spreading out to the Hampshire Downs and Wealden Heights, which form the limits of the basin to the south.

I. To the south of Suffolk lies the county of **Essex,** so named after its first conquerors, a tribe which took the name of the *East Saxons.* It is the largest county of the Thames group, having an area of 1,657 square miles, and forms a rough square bounded by the German Ocean on the east, and by rivers on its three other sides. The *Stour* parts it from Suffolk on the north, the *Lea* and the *Stort* sever it from Middlesex and Hertfordshire on the east, and the *Thames* cuts it off from Kent on the south. Its physical structure is the same as that of the East Anglian counties; but the chalk which forms so great a feature in these only passes over the north-western corner of Essex on its way to Hertfordshire, while the strip of alluvial soil which edges the eastern counties becomes in Essex a broad tract of clay extending over the greater part of its surface. It is however only along the sea-shore and along the banks of the Thames that these clays show themselves in flat ground, and form a fringe of low unhealthy marsh, where fever and ague have been but partially driven out by drainage. In the bulk of the county, but especially on its western side, the Essex clays rise into low hilly ground once covered by forests such as those of *Epping* and *Hainault,* which have now given place for the most part to rich meadow-lands.

The chief rivers within the county take their rise in the chalk downs of the north-west, and wind among the clay hills eastward to the German Ocean, as the *Colne*, the *Blackwater*, and the *Crouch*, or southward to the Thames, as the *Roding*. The coast is deeply indented by shallow creeks and bays formed at the river mouths, and broken into a number of islands which are little more than tracts of low swamp. The industry of Essex is almost wholly agricultural, though some little manufacture of silk survives among the small towns, and a fishing-trade is carried on in the ports along the coast. Its population, 466,436, is larger than that of Norfolk, though the county is so much smaller : and is very evenly distributed over its whole surface in small towns scattered along the various river-valleys. *Harwich* is a little fishing town on the mouth of the Stour; on the Colne lie *Halstead* and *Colchester*. This last is the only large town of the county, and its prosperity depends partly on its great markets for corn and cattle, and partly on its valuable oyster-fisheries. The Blackwater, which is known in the upper part of its course as the Pant, has near its source the little town of *Saffron Walden* lying in the midst of chalk downs, and surrounded by fields of saffron whence the London market is supplied ; lower down the stream is *Braintree*, and near its mouth the fishing port of *Maldon*. *Chelmsford*, the county town, lies on the *Chelmer*, an affluent of the Blackwater. At the mouth of the Thames is *Shoeburyness*, the chief artillery station in

England. The little town of *Barking* lies on the
Roding as it enters the Thames.

II. The tract of alluvial clay which covers the bulk
of Essex stretches westward along the Thames, whose
ancient river-bed it represents, to form the smallest
but one of English counties, **Middlesex.** This
county, which takes its name from the tribe of Saxons
who settled this *middle* district, has an area of 282
square miles, and a very irregular form determined by
its river boundaries. The chalk downs of Hertfordshire
run along its northern borders ; but on the south the
windings of the *Thames* part it from Surrey ; on the
east the *Lea* flows between it and Essex ; on the west
the *Colne* severs it from Buckinghamshire. Another
small river, the *Brent*, which like the rest joins the
Thames, crosses the centre of the county, where it
winds among the low clay hills which vary the surface
of Middlesex. The rising ground of *Harrow* lies to
westward of its waters, that of *Hampstead* and *Highgate* to eastward.

Middlesex, though so very insignificant in point of
size, is yet the most important of English counties.
Its population, 2,539,765, is larger than that of
Yorkshire, a shire which is 21 times greater in mere
extent, and at the same time one of the most
thickly peopled in England. This importance is due
to the fact that within the limits of Middlesex lies the
chief part of the wealthiest and most populous city
in the world, LONDON, the capital of the British
Empire. The greatness of this city is the consequence
of the advantages of its geographical position. Situ-

ated at the outlet of the valley of the Thames, it is
enriched by the agricultural wealth of the most fertile
district in Britain ; lying at the centre of the network
of roads which converge from every part of England
on the chief sea-port of the country, it forms the
trading centre of the island ; its site on the magnificent estuary of the Thames, with its relation to the
neighbouring seas and the European mainland, make
it the first commercial city of Europe ; and finally its
position at the centre of the continental masses of the
globe constitutes it the greatest among all international
trading ports.   Since the importance of London thus
depends mainly on its commercial greatness, it only
attained its true development with the rapid growth of
commerce which followed on improved means of
navigation in the last century.   As the capital of
international trade the city has drawn into itself since
that time a larger number of inhabitants than have
ever before been gathered into one place—3,250,000
persons ; that is, a greater mass of people than that
contained in any county of England, even in Yorkshire or Lancashire.   This population is still growing
at so rapid a rate that the increase is equal to the
addition of a village of 240 persons in every day of
the year, or the addition of more than one house to
the existing mass in every hour ; and the length of
the streets is already so great, that if stretched out in a
continuous straight line they would reach across Europe
and Asia, as far as the southern borders of India.   A
part of this vast city extends to the southern banks of
the Thames, and lies therefore outside the borders of

Middlesex, so that its population is only partially included in that shire.

The greatness of London reduces the remaining towns of the county to places of little importance. They are *Brentford*, the county town, at the junction of the Brent and the Thames; *Hounslow* on the borders of a great heath to westward of it; and *Harrow*, famous for one of the great schools of England.

III. The chalk uplands which form so small a portion of Essex play a much larger part in the county to which they pass on the west, **Hertfordshire**. This shire has an area of 611 square miles. It does not touch the Thames, though lying within its basin, but belongs wholly to the range of the Chiltern Hills which enter it from Buckinghamshire on the south west, and stretch out beyond it to Cambridgeshire on the north-east. The steep escarpment of this range runs along the north-western side of the county and overlooks the plain of Bedford and the valley of the Ouse; but the bulk of Hertfordshire consists of the long gentle slopes by which its uplands fall to the low country of Middlesex and Essex on the south-east.

The rivers of the shire, as the *Colne* and the *Lea*, are thrown down these slopes to the valley of the Thames; and along their banks lie quiet little agricultural towns, with a small industry in straw-plaiting. The chief among them are *Hertford* the county town, and *Ware* on the Lea, and *Bishop Stortford* on the Stort, one of its tributaries. In the valley of the

Colne are *S. Albans* and *Watford: Hitchin*, a town
larger than any of these, with straw-plait manufactures, lies on the north-western border.   The population of the whole shire is 192,226.

IV. From Hertfordshire the line of the Chilterns
stretches to the south-west across the central part of
the county of **Buckingham**.   This shire is a long
narrow strip of land with an area of 730 square miles,
ranging north and south between the valleys of the
Ouse and of the Thames.   The Ouse parts it from
Northampton, the Thames from Berkshire; on the
east lie Middlesex, Hertfordshire, and Bedfordshire,
while Oxfordshire extends along the whole of its
western border.

The chalk downs of the Chilterns, which cross the
southern district of Buckinghamshire, and are parted
from Middlesex by the Colne, form its smallest and
least important part, and till modern times were little
more than a wild forest country.   As they slope
rapidly to the valley of the Thames, several towns
gather on or near the river-banks, *High Wycombe*,
Great Marlow, and *Eton*, famous for containing the
largest school in England.

The northern and larger part of Buckinghamshire
lies in the great plain which is overlooked by the
steep escarpment of the Chilterns.   The *Vale of
Aylesbury*, which lies directly below the chalk cliffs, is
one of the richest grazing lands in England, and is
watered by the *Thame*, an affluent of the Thames,
which has on its banks the county town of *Aylesbury*.
In this district we have entered on the upper basin of

the Thames, that is to say, on that part of it which is shut in between the oolitic and chalk ranges ; and from this we pass, in the extreme north of the county into the basin of the *Ouse*, where the plain sends its waters, not to the Thames, but to the Ouse, which just crosses its northern border. On this river lie the little towns of *Buckingham* and *Olney*. The industry of Buckinghamshire is chiefly agricultural, but to this is added some small manufactures of straw-plait, paper, lace, boots, and shoes. The population, 175,880, is not great for its size, being less than that of Hertford, a smaller county.

V. As we pass from Buckinghamshire westward, whether we follow the course of the Chiltern Hills or of the valley of the Thame, we are led directly into **Oxfordshire**, through which both the uplands and the plain are continued. This county is of almost exactly the same size as Buckinghamshire, having an area of 739 square miles, and is very irregular in form. It consists of a broad mass of country which stretches along the northern banks of the Thames from Gloucestershire to Buckinghamshire, and sends out a tongue of land by the eastern bank of the same river as it bends southwards between the towns of Oxford and Reading, as far as the point where it breaks through the chalk range into its lower valley. The shire thus lies wholly within the upper valley of the Thames (see p. 59).

The broad northern portion of the county is made up of that low and broken belt of oolitic uplands which connects the Cotswolds on the west with the Northampton uplands on the north-east, and forms along the north-western side of Oxfordshire the escarpment of the *Edge Hills*, overlooking the plain of Warwickshire and the valley of the Avon. Small rivers are thrown down from the crest of these uplands to the valley of the Thames at their base, as the *Windrush*, the *Evenlode*, and the *Cherwell*, and along these streams and their tributaries lie the quiet little agricultural towns of this district, such as *Witney ; Chipping Norton* and *Woodstock ;* and *Banbury*. This last town lies on the Cherwell as it enters the county from the north ; on the banks of the same river as it passes southward into a low and perfectly level tract of land where it empties itself into the Thames, is the chief town of the shire, *Oxford*, the seat of a bishopric, and of one of the two greater English Universities.

Close to Oxford the Thames bends sharply to the south, and in this part of its course it passes through country of a different character. The narrow neck of land which runs along the river between Berkshire and Buckinghamshire, and is traversed by the lower valley of the *Thame*, forms a belt of low flat ground dotted with agricultural villages, but without a single town of any size. This plain is bounded to the south-east by the escarpment of the *Chiltern Hills*, which strike across the extreme southern part of Oxfordshire to the very borders of the Thames, where the river has cleft a channel through their soft rocks and bends round their extremity in a half-circle to *Henley*, the only important town of southern Oxfordshire. The whole population of the shire, which is chiefly gathered in its northern half, is about 178,000.

VI. In Oxfordshire we have been brought very near the head-waters of the Thames—we now therefore cross the river to travel along the counties of its southern bank. The great chalk range whose course we have followed from Norfolk to Oxfordshire has led us to the left bank of the Thames, at the point where that river breaks its way southwards through the Chiltern Hills (see p. 59). On the opposite or right bank of the stream the chalk range again rises in the escarpment of the *Ilsley Downs* and *White Horse Hill*, which traverse the county of **Berkshire** from the western banks of the Thames to the Marlborough Downs of Wiltshire.

By this line of heights Berkshire is broken into two distinct parts. (1.) The steep white cliffs which form the northern face of the range look out over the *Vale of White Horse* and the valley of the river *Ock*, which runs from west to east below the heights; this valley belongs to the *upper* basin of the Thames. (2.) To the south the downs fall in gentle slopes to the valley of the *Kennet*, a river which also flows from west to east, parallel to the line of the chalk uplands; but whose course lies within the *lower* basin of the Thames. On the further side of the Kennet valley rises the steep escarpment of the Hampshire Downs, which close in Berkshire to the south by a long straight line of heights. The shire thus consists of a square mass of country lying in the upper valley of the Thames between Wiltshire and

Oxfordshire, with a narrow strip of land thrown out eastward into the lower valley of the same river between Hampshire, Surrey, and Buckinghamshire. In fact Berkshire is in form not at all unlike an inverted Oxfordshire, and is very nearly the same size, having an area of 705 square miles.

Its industry is wholly agricultural. In its northern part, in the fruitful valley of the Ock, are scattered quiet little country towns, *Farringdon* and *Wantage*, with *Abingdon* and *Wallingford* on the Thames. To the south of the chalk downs *Hungerford* and *Newbury* lie on the banks of the Kennet; and at the place where this river pours its waters into the Thames is the large county town, *Reading*. It is at this point that the Thames, having finally broken through the chalk uplands, enters fully on its lower basin, and having taken up its first tributary, the Kennet, turns eastward by the neck of low clay land sent out in this direction from the main body of Berkshire, and passes under *Maidenhead* and the Royal palace of *Windsor Castle*. Berkshire is more thickly peopled than Oxfordshire, having in a smaller tract of country 196,475 inhabitants.

VII. As we travel along the course of the Thames eastward from Berkshire, and enter a much wider part of its basin, we pass into the county of **Surrey**, so-called from the *south rik*, or kingdom, of the Saxons. This is a county rather larger than Berkshire, with an area of 748 square miles; it forms a rough square which stretches from the southern bank of the Thames, as it flows past Middlesex, to the Wealden

Heights of Sussex, which shut in the Thames basin to the south. The Hampshire Downs run in a sharply marked escarpment along its western border, and send out eastward the range of the North Downs, which crosses the centre of Surrey and passes into Kent on its eastern side.

The chalk uplands of the North Downs form the chief feature of the scenery of the shire. Beginning in the narrow line of the *Hog's Back*, near Farnham, a perfectly straight ridge of six miles in length, they gradually widen to the downs of *Epsom* and *Croydon* in the east. On their way they are cleft in several places by natural gaps, and by the channels of *rivers* such as the *Wey* and the *Mole*, which have their head-waters in the Wealden Heights, and cut through the chalk downs on their way to join the Thames. The passes thus formed are of very great importance, since it is they that determine the lines along which the roads from London to the south must pass. These breaks in the line of the North Downs are therefore marked by the chief towns of the county, which lie so as to command the passes : *Farnham* between the Hampshire Downs and the eastern extremity of the Hog's Back ; *Guildford*, the county town, situated on the Wey as it cleaves its passage through the chalk heights ; *Dorking* in a similar pass formed by the Mole ; and *Reigate* near a pass to eastward of it, noted for its mineral springs.

The Downs fall on their southern side to a narrow belt of low undulating country belonging to the Weald, and composed of heavy clay. Its soil fur-

nishes a fuller's earth which gives rise to a small
industry, but for the most part its villages are wholly
employed in agriculture. The northern escarpment
of the Downs overlooks the low clay valley of the
Thames; here the population gathers more and more
thickly as we near the river banks. Two large districts
of London, *Southwark* and *Lambeth*, lie on the Surrey
side of the Thames, and the country for miles round
forms a succession of towns which are simply suburbs
of London. *Richmond* and *Kingston* are situated on
the river itself. The population of Surrey is very
large, nearly 1,100,000, the greater part of which
belongs to the thickly-peopled districts about London
and its suburbs.

VIII. If we follow the line either of the Thames
valley, or of the North Downs as they pass eastward
beyond the border of Surrey, we are led directly into
Kent, a county which forms the south-eastern extremity of England. With an area of 1,627 square
miles, it is nearly as large as Essex, which lies just
opposite to it at the mouth of the Thames, and more
than twice as large as any other county of the Thames
basin. From the river estuary on the north it extends
southwards to the Wealden Heights of Sussex, the
southern boundary of the Thames basin; and the
Straits of Dover and the North Sea lie along its
eastern shores.

The inner character of Kent is the same in
all its features as that of Surrey. The North Downs,
which enter the county from the east, widen constantly as they pass eastward to the sea, till they

end in the chalk cliffs of the *Isle of Thanet*, at the mouth of the Stour, and the broad mass of uplands which stretch from thence southward along the Straits of Dover to *Folkestone*. These chalk heights therefore occupy the main portion of the county, but to the south of them lies, as in Surrey, a belt of undulating country belonging to the Wealden district. This lower land is separated from Sussex by the river *Rother*, and ends in the English Channel in *Romney Marsh*. Under the northern face of the Downs, and parting them from the Thames, runs a narrow strip of low clay, broken by river estuaries and marshy islands, such as the *Isle of Sheppey*, at the mouth of the Medway.

The line of the North Downs is cleft, as in Surrey, by the valleys of rivers flowing northwards to the Thames; the little valley of the *Darent* to the west, the much greater *Medway* in the centre, and the *Stour* in the east, which falls into the German Ocean after passing round the Isle of Thanet. This river was formerly connected with the Thames estuary by a channel of the sea opening at Reculvers and cutting off Thanet into an island, but the channel is now dried up, leaving Thanet united to the mainland. The chief towns of Kent lie along the roads which follow these rivers in their passage through the hills. The Medway rises near the mineral springs of *Tunbridge Wells*, on the borders of Sussex, and winds among low clay hills past *Tunbridge; Maidstone*, the county town, marks the place where its channel cuts through the chalk downs; and *Rochester*, an old cathedral city, guards the opening of the pass where the river emerges on the low clay valley of the Thames. The passage of the *Stour* in its long deep valley across the Downs is marked by *Canterbury*, a very ancient cathedral city, and the seat of the first Archbishop of England. The whole of the country traversed by these rivers is very fertile, abounding in cherry-orchards and great hop-gardens: the hop-grounds alone cover 50,000 acres, and their cultivation affords the chief industry of the agricultural districts.

But the inland towns are few, as in all agricultural counties, and the larger part of the population of Kent lies along its shores to north and east. The chalk cliffs in which the North Downs abruptly end to eastward on the sea are marked by summer bathing-places, such as *Margate* and *Ramsgate ;* and by seaport towns such as *Deal, Dover*, and *Folkestone*, lying on the Straits of Dover. The most wealthy and important trade of Kent, however, is that which gathers in its northern parts, along the banks of the Thames, where the river may still be considered as the vast port of London, and where its enormous commercial and military importance have given rise to an almost unbroken succession of busy towns and great dockyards. Not far from London are the docks of *Deptford*, and the naval hospital of *Greenwich*, and beyond these the chief military arsenal of England, *Woolwich*, each of these towns having a population of 40,000 inhabitants. The shipping-port of *Gravesend* is smaller, but *Chatham*, on the estuary of the Medway, with its immense docks and naval arsenal, has 44,000 inhabitants ; and at the mouth of the same river, on the Isle of Sheppey, are the dockyards of *Sheerness*. This side of the Thames estuary, with its succession of docks crowded by vessels from every country of the world, forms a striking contrast to the silent and deserted marshes of its opposite banks lying within the county of Essex. The important shipping industries of Kent have raised its population to 848,300, a number nearly double that of Essex.

CHAPTER XVI.

THE basin of the Thames is bounded on the south by a line of low heights, the Wealden Heights, the Hampshire Downs, and Salisbury Plain. These uplands form the *southern water-parting* of England, from whence a group of small rivers is thrown down to the English Channel (see p. 21). The district which lies between this water-parting and the sea is occupied by four counties—*Sussex, Hampshire, Wiltshire,* and *Dorsetshire.* The whole group forms a long and narrow triangle laid along the sea-coast; Sussex forms its eastern point, while its broad western side is made up of Wiltshire and Dorsetshire. (See p. 108.)

These counties, lying as they do to the south, and open to soft southerly winds, are very much milder in climate than the more northerly parts of England. The harvest in Hampshire is a fortnight earlier than it is in Yorkshire, and fruits and shrubs which will not bear the climate of central England can be grown in its southern shires.

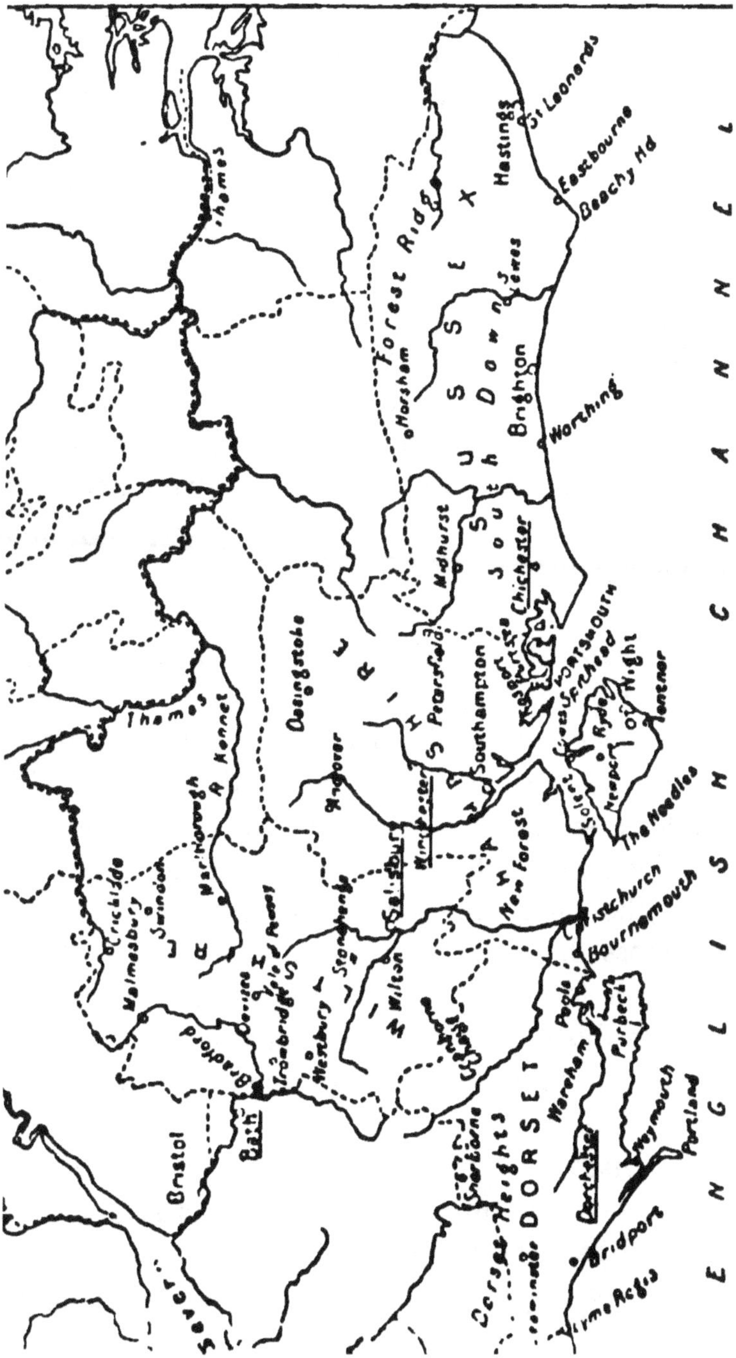
ENGLISH CHANNEL
SUSSEX
DORSET
South Downs
Forest Ridge
St Leonards
Hastings
Eastbourne
Beachy Hd
Lewes
Horsham
Brighton
Worthing
Midhurst
Chichester
Petersfield
Portsmouth
Spithead
Southampton
Ryde
Newport
Ventnor
The Needles
Solent
Winchester
Andover
Basingstoke
Salisbury
New Forest
Christchurch
Bournemouth
Poole
Wareham
Purbeck
Weymouth
Portland
Dorchester
Bridport
Lyme Regis
Sherborne
Bristol
Bath
Bradford
Trowbridge
Devizes
Malmesbury
Swindon
Marlborough
Cricklade
Thames
R. Kennet
Wilton
Stonehenge
Severn

I. To the south of Surrey and Kent a narrow belt of country stretches along the shores of the English Channel from Romney Marsh to the Hampshire Downs. This is **Sussex**, the settlement of the *South Saxons* in early times. Its area is 1,460 square miles, and its scenery throughout is of a very broken and varied character. The county naturally divides itself into three parallel belts of country, the Weald, the South Downs, and the coast. (1.) Along the northern limits of the shire, where it borders on Kent and Surrey, runs from west to east the broken line of the *Wealden Heights*, or the *Forest Ridge*, whose southern slopes covered with woodland fall to a deep clay basin, also belonging to the district of the Weald. (2.) From the further side of this valley rises the steep escarpment of the *South Downs*, which come from the Downs of Hampshire in the west, and pass eastward to the sea at *Beachy Head* in a long level line of heights, with the soft monotonous outline peculiar to chalk hills. (3.) The straight lines of the Downs are only broken here and there by narrow channels cut by the little Wealden *rivers*, the *Ouse*, and the *Arun*, as they cleave their way across the chalk, and enter on the strip of alluvial soil which forms the low sea-coast. To westward of the Arun this strip widens to a broader tract of clay, broken by creeks and inlets of the sea, of which the chief is *Chichester Harbour*.

The district of the Weald in the north of Sussex was till modern times one of the wildest and least inhabited parts of England. The heights which form the water-parting were covered with the dense masses of forest from which they took their name, and were traversed only by narrow lanes or mule-tracks. The country is now brought under cultivation, but is still thinly peopled, and has only two small towns, *Horsham*, and *Midhurst*. The inhabitants of its villages live by agriculture, hop-picking, foresting, and the making of barrels from the wood grown on their hill-sides.

The greater part of the population of Sussex is therefore gathered in the towns of the coast district. *Lewes* lies in the gap made by the river Ouse, through the South Downs, and commands the high road from London to the sea-ports of the coast, which in old times were some of the most important in England. Since the use of vessels too large for their harbours, however, they have sunk into mere fishing-ports, or watering-places. The chief of these are *Hastings*, and near it *St. Leonard's; Eastbourne; Brighton*, with 90,000 inhabitants, the largest watering-place in England; *Worthing;* and *Chichester*, at the head of Chichester Harbour, a town which has lost its old greatness, and is now only important as being the county town and a cathedral city. The population of Sussex is 417,460.

II. If we follow the line of water-parting westward from Sussex, we pass at once into the county of **Hampshire**, where the sources of the rivers lie along the crest of a great mass of chalk uplands known as the *Hampshire Downs*. These downs form a steep escarpment which runs along the northern border of the county, and from the summit of which we may look southwards over the whole

extent of Hampshire, with its area of 1,672 square
miles. The Downs on which we stand are prolonged
across the western border of the county into Wiltshire,
while on its eastern side they break up into the North
Downs of Surrey, and the South Downs of Sussex.
To the north their escarpment, curved into bays and
headlands, overlooks the valley of the Kennet in
Berkshire ; and on the south they fall in long undulating lines to a level tract of alluvial soil which
borders the shores of the English Channel. On this
side they throw down small rivers, such as the *Test*
and the *Itchin*, which traverse the low ground that
lies at their base, and finally empty themselves into
the Southampton estuary. This tract of chalk uplands, with its great reaches of gently swelling downs,
scantily wooded, and scarred with patches of white
soil gleaming here and there through its grasslands, is
wholly given up to agriculture. Quiet little villages and
towns lie in the hollows where its undulations dip
and rise again, or in the sheltered river valleys. Near
the centre of the shire is the county town *Winchester*,
which was one of the earliest and most important
cathedral cities of England : the little towns of *Andover* and *Basingstoke* lie in the north, and that of
Petersfield on the eastern border of the downs.

But though the chalk downs cover so great an
extent of ground, they do not form the most important part of Hampshire, for it is the belt of alluvial
soil by the coast which contains the chief wealth and
trade of the county. The strip of clay which borders
the sea from Chichester Harbour to the Test is broken

by two of the greatest harbours in England. To the east lies *Portsmouth Harbour*, with its immense dock-yards and naval arsenal, and clustered round it a group of towns, *Portsmouth* having 113,000 inhabit-ants, *Portsea*, *Kingston*, and *Gosport*. In the middle of the coast is the great estuary of *Southampton Water*, and here the large town of *Southampton*, at the head of the estuary, with 54,000 inhabitants, is the centre of a very busy foreign trade, chiefly with the Mediterranean.

On the western shores of Southampton Water the lowlands broaden into a square tract of country which stretches out to the borders of Dorsetshire, and forms the district of the *New Forest*, formerly the hunting-ground of the English Kings, and still in great part covered with wood, and without any towns (p. 59). It is bounded on its western side by the river *Avon*, at whose mouth lies the seaport of *Christchurch*, and near it the sheltered watering-place of *Bourne-mouth*. Hampshire contains altogether 544,684 inhabitants.

The Isle of Wight is geographically a part of Hampshire, though cut off from it by the channels of the *Solent* and *Spithead*. Its northern half is formed of a tract of low clay like that of the neighbouring coast of Hampshire; its southern half consists of a line of chalk heights, which are continued eastward from the peninsula of Purbeck, and run across the island from the sharp rocks called the *Needles* to the *Culver Cliffs*. The Isle of Wight is famous for its beautiful scenery, and for the mildness of its climate, in which the arbutis, the grape, and the fig will flourish. The chief towns are *Newport*, having near it the ruins of *Carisbrook Castle;* *Cowes* and *Ryde*, watering-places on the northern coast; and *Ventnor*, which lies sheltered under the chalk cliffs, or *Undercliff*, of the southern shore.

III. If we follow the water-parting along the crest of the north Hampshire Downs, we are led westward into the very midst of **Wiltshire**, where the division between the rivers is continued along the summit of an escarpment of chalk which strikes across the centre of the county. This escarpment divides Wiltshire into two parts.

(1.) It forms the northern edge of a great tableland of undulating chalk country known as *Salisbury Plain* —a tableland twenty miles broad and fifteen miles long, which consists of a tract of barren and woodless country, raised 500 or 600 feet above the sea, covered with a short thin grass, and with a soil which is only by degrees being brought under cultivation. Its population is very scanty, and its few towns gather in the lower valleys of its streams, where the climate is more sheltered and the soil more fertile. The chief river is the *Avon*, which flows southwards across the downs; its tributary, the *Wiley*, comes from the west. *Salisbury*, the county town and a cathedral city, lies on the Avon; near it is *Stonehenge*, a very ancient circle of great stones whose origin is unknown; *Wilton*, with carpet manufactures, is situated on the Wiley. These downs of southern Wiltshire are bounded on the west by the oolitic uplands of Somerset; but on the south they extend into Dorsetshire, and on the east into Hampshire, where they break up into the chalk ranges of south-eastern England.

(2.) The escarpment which forms the northern edge of Salisbury Plain overlooks to the northward the *Vale of Pewsey*, a narrow tract of fertile country which runs across the country from west to east, and contains the agricultural town of *Devizes*. Beyond this valley lies that half of Wiltshire which stretches northward to the Thames, and lies (*a*) partly within the basin of the Thames, and (*b*) partly within the basin of the Severn.

(*a*.) The northern side of the Vale of Pewsey is, like the southern, bounded by an escarpment of chalk, which forms the edge of a second upland plain, the *Marlborough Downs*. This tract of chalk is of the same character as the first, barren and treeless. It extends eastward to Berkshire, and its northern face overlooks the *valley of the Thames*, part of which, from *Swindon* to *Cricklade*, lies within Wiltshire. A deep depression across the centre of these Marlborough Downs forms the upper valley of the *Kennet*, a tribu· tary of the Thames, which has on its banks the only town of this district, *Marlborough*.

(*b*.) Under the western escarpment of the Marlborough Downs lies the *valley of the Avon*, a river which empties itself at Bristol into the estuary of the Severn. It flows among low oolitic clays, which rise on the west into the Cotswold Hills of Gloucestershire. The towns along its valley, situated as they are close to the Gloucestershire border, share for the most part in the clothing manufacture which gathers round the Cotswold Hills. They are *Malmesbury* in the north ; *Bradford* and *Trowbridge* near the Avon as it crosses the opening of the Vale of Pewsey, with *Westbury* at a little distance.

The form of Wiltshire is an oblong square, like that of Hampshire, but its area is less, being 1,352 square miles. Owing to its great tracts of barren soil the whole county has a population of only 257,177 persons, or not quite so many as are contained in the single town of Leeds.

IV. The Downs of Wiltshire extend across its southern border into **Dorsetshire,** the smallest county of the southern group. It forms a rough triangle with an area of 987 square miles, which extends along the Channel with its broad eastern end resting on Hampshire, and its western point thrust between the shires of Somerset and Devon.

Dorsetshire is remarkable as forming the point where all the upland ranges of eastern England are gathered together into one stem, from which they branch out, as we have seen, over the whole surface of the country (see p. 59). Both the chalk downs of Wiltshire, and the oolitic uplands of Somerset are carried within its borders, and run southwards in parallel lines, only separated by the *Vale of Black-more*, to end side by side on the sea coast near Lyme Regis. The *Dorset Heights*, the most important of these two ranges, form a chain of chalk uplands which extend from Salisbury Plain by *Cranborne Chase*

to Lyme Regis, in a curved line of hills which reach
their greatest height of 900 feet near *Beaminster.*
From this point they throw out eastward by the sea-
coast the long range of the *Purbeck Heights,* which
connects *Purbeck Island* with the mainland.   The
lesser range of *oolitic uplands* extends from Somersetshire past the little town of *Sherborne* as far as
Lyme Regis; and from here these uplands also
send out eastward along the coast past *Bridport,*
a long low line of oolitic clays which end in the
Isle of Portland, a small peninsula connected
with the mainland by *Chesil Bank,* a ridge of loose
pebbles ten miles long. On the sheltered bay
shut in between the Isle of Portland and the Purbeck Heights is situated a large watering-place,
Weymouth.

The only low-lying country of Dorsetshire is its
eastern portion, which is inclosed between the Purbeck
Heights and Cranborne Chase, and which adjoins
Hampshire on the east.   This district, composed
of heavy clay like that of the New Forest, is traversed
by two rivers, the *Stour* and the *Frome.* The Stour
as it enters the county from the north, crosses the
Vale of Blackmore, cuts a way for, itself through
the Dorset Heights, and finally passes into Hampshire, where it joins the Avon. The Frome, which
rises in the chalk hills by Beaminster, flows in a deep
valley below the Purbeck Heights for their whole
length, and falls into *Poole Harbour.*

The county town, *Dorchester,* lies on the banks of
this river; and lower down the stream is *Wareham,*

close to Poole Harbour. *Poole* is a seaport on the western banks of the harbour : the clay in its neighbourhood is much worked for the Staffordshire potteries, and is exported through Poole.

The chief industry of Dorsetshire is agricultural ; its population is 195,537.

CHAPTER XVII.

WE have now followed the course of the southern water-parting in a tolerably direct line from its eastern extremity in the Wealden Heights to its junction with the main water-parting of England in the Wiltshire Downs. (See p. 21.) From this point its course becomes extremely irregular, forming a sinuous line which extends along the peninsula of the south-west; and streams are sent down from it to the channels on either side, to north-west and to south-east. The country drained by these streams is occupied by three counties, *Somersetshire, Devonshire*, and *Cornwall* (p. 108). Somersetshire forms the basin of those streams which empty themselves into the Bristol Channel; Devonshire and Cornwall are drained by rivers flowing to the Atlantic Ocean and the English Channel. These three counties are also linked together by the fact that they belong to the mountain group of the south-west, which is mainly built up of very ancient rocks, and contains traces of an ancient population. They share in the same mild and damp climate, a

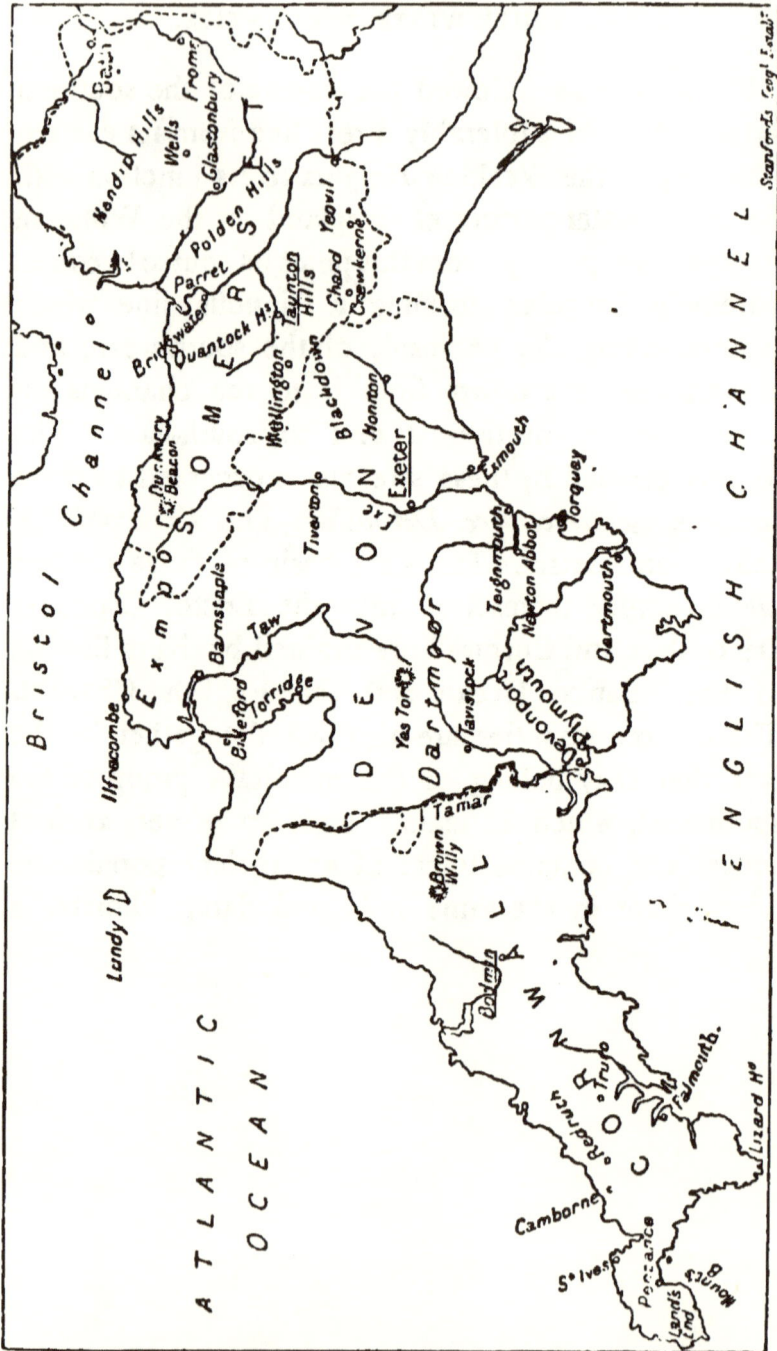

SOUTH WESTERN COUNTIES.—Towns and Rivers.

climate common to counties lying so far south, and almost surrounded by sea.

I. The water-parting of southern England passes from Wiltshire westward in an uneven line which exactly marks the western and southern boundaries of the county of Somerset. Its northern border is formed by the river Avon, a river thrown down almost from the meeting-point of the two water-partings. This shire therefore exactly forms the basin of those streams which fall from the southern water-parting into the Bristol Channel.

Somersetshire extends very nearly across the north-eastern extremity of the peninsula included between the Bristol and the English Channels, being only separated from the sea on the south by a narrow part of Dorsetshire. It has an area of 1,636 square miles, and resembles in form an arm bent at right angles round the eastern and southern shores of the Bristol Channel. The valley of the river *Parret* crosses it at the bend of the elbow, and divides the county into two very distinct parts—the broader part runs northwards between Wiltshire and the Severn estuary to Gloucestershire ; the narrower part is thrust out westward like a wedge between Devonshire and the Bristol Channel. These two districts differ from one another geographically in the structure of their rocks and in the character of their scenery, and politically in their population and history.

(1.) That part of Somersetshire which lies to the north of the Parret consists of a very broken country of hill-ranges and low marshes. Its eastern border is marked by a line of very irregular oolitic uplands which stretch from *Crewkerne,* on the borders of Dorsetshire, to *Bath,* where they rise into the Cotswold Hills and pass into Gloucestershire. The town of *Frome* among these uplands carries on the clothing manufacture common in the neighbourhood of the Cotswolds. From these uplands of the eastern border two parallel ranges of heights, the *Mendip* and the *Polden* Hills, are thrown out westward to the Bristol Channel. The Mendip Hills form the most important range; they rise at their greatest height to 1,100 feet, and their top forms a flat with a rapid slope on either side ; their rocks of mountain-limestone contain veins of lead, copper, and other metals. The lines of hills are bounded on either side by low river valleys. (*a.*) To the north of the Mendip Hills lies the lower valley of the *Avon,* which forms the boundary between Somersetshire and Gloucestershire. As the river enters the county it has on its banks the cathedral city of *Bath,* with 52,000 inhabitants, famous from very early times for its hot mineral springs. In the district to the south of the river lies a part of the Bristol coal-field which was once united with the great coal-measures of South Wales. (*b.*) To the south of the Mendip Hills and parting them from the Polden Hills, is the sunken ground of the *Brent Marshes* drained by the river *Brue ;* these marshes form an immense tract of swamp containing peat, which is cut for fuel. The Brue passes by *Glastonbury,* now a very small town, but in old times the most famous monastery in England. Not far from it is *Wells,* little larger in size, but the seat of a bishopric. (*c*) The *Parret* which flows to the south of the Polden Hills receives two small tributaries, the *Isle*, which rises by the little town of *Chard*, and the *Yeo* or *Ivel*, which has on its banks the large town of *Yeovil* with glove manufactures. The Parret falls into Bridgwater Bay close to the Brue ; the seaport of *Bridgwater* on its estuary has a considerable shipping-trade.

The low grounds of these river valleys are chiefly given up to pasture, and large quanties of butter and cheese are made in this part of the county.

(2.) If we now cross the Parret we enter on a second and smaller part of Somersetshire, extending to the west. This division consists almost wholly of a mass of mountainous country belonging to the ancient rocks of the older mountain groups. A part of the high moorlands of *Exmoor* extends from north Devon so as to fill the whole western extremity of Somerset, rising in *Dunkerry Beacon* to a height of 1,700 feet, and in *Brendon Hill* to 1,300 feet. To the east of these is the range of the *Quantock Hills*, rising in parts to 1,300 feet, and overlooking the valley of the Parret. Opposite to them, on the southern border of the county, lie the *Blackdown Hills*. The whole of this mountainous district is wild and barren, with scarcely a village to break the solitude of its great tracts of bog and moorland heather. The only low ground is the deep *Vale of Taunton*, formed by the river *Tone* as it flows between the Quantock and Blackdown Hills to the Parret ; in the midst of this fruitful valley lie the small town of

P

Wellington, and the large county town, *Taunton*. The whole population of Somerset is 463,483, the great bulk of which lies in its northern half.

II. **Devonshire**, lying between the counties of Somerset and Cornwall, stretches across the broadest part of the peninsula from the Bristol to the English Channel, and thus occupies that portion of the south-western mountain-group where the moorlands rise to their greatest height and present their grandest and wildest scenery. This shire is the third in size among English counties, having an area of 2,589 square miles. It naturally divides itself into three distinct parts. (1) *Exmoor* in the north, (2) *Dartmoor* in the south, and (3) a broad plain of pasture-land which lies between them.

(1.) Exmoor, the smallest and least important division, consists of a high table-land covered with mountain pasture and great tracts of bog, and absolutely treeless save in the little river-valleys. A number of small rivers are thrown down from these heights southward to the plain, the chief of which is the *Exe*. The whole district is almost uninhabited. There is but one small town, *Ilfracombe*, which lies on the cliffs at the western extremity of the Exmoor heights overlooking the sea, and is famous for the beauty of its scenery.

(2.) *Dartmoor* in south Devon, severed from Cornwall by the valley of the *Tamar*, is a high table-land like that of Exmoor, but steeper, wilder, and more extensive. Its length from north to south is twenty-two miles, and its breadth from east to west fourteen miles: it forms the highest land in England to the south of the Trent, rising above the sea-level as far as 2,000 feet in *Yes Tor*, while its general height is about 1,200 feet. Its surface is barren, with little grass and great reaches of bog; but the rocks are rich in mines of lead, iron, tin, and copper, and in granite, limestone, and veined marble for building purposes. A group of head-waters is formed in the high district near Yes Tor, from whence rivers are thrown down in every direction, and open estuaries in the coast, by which the mineral wealth of the country finds an outlet to the sea. Some of these streams have a short, rapid course to the English Channel on the south and east, as the *Tavy*, the *Plym*, the *Dart*, the *Teign*, and many more. The *Tawe*, on the other hand, falls down to the west, and has a longer course to Bideford Bay. The little river-valleys of Dartmoor are marked by a number of towns which cluster round the foot of the heights, and are mostly engaged in the mining and export trade of the district, while some of them are fashionable watering-places. The most important among these towns are *Plymouth*, a great seaport and naval arsenal, with 68,000 inhabitants, on *Plymouth Sound*, at the mouth of the Plym; and *Devonport* on the Tamar, with dockyards and naval arsenal, containing 50,000 people. Lying close together and having the same industries, these have gradually grown into one town, and by their junction have formed the most populous city on the southern coast. *Tavistock* has grown up round the mines of the valley of the Tavy; the other towns lie on the river estuaries, such as

Dartmouth on the Dart; *Newton Abbot* and *Teign-mouth* on the Teign; and *Torquay* on the north side of *Tor Bay*, a large watering-place. Dartmoor and the district immediately round it is thus the most important and populous part of Devonshire, being the centre of all its mining and shipping industries.

(3.) Between Dartmoor and Exmoor stretches an undulating plain, which rises into low hills on the north-western coast round Bideford Bay. It is watered by the *Exe* from Exmoor, and the *Tawe* from Dartmoor, and has besides its own group of head-waters in the hills of the extreme west, from which the *Tamar* is thrown down to the English Channel, and the *Torridge* to Bideford Bay. The plain forms some of the richest pasture-land in England, which is divided into large dairy farms. Its towns lie in the river-valleys. On the Exe are *Tiverton*, and *Exeter*, the county town and the seat of a bishopric; *Honiton*, with its lace manufactures, lies on a little stream near the eastern border of the shire; in the north-west, at the mouth of the Tawe is the port of *Barnstaple;* at the mouth of the Torridge the smaller port of *Bideford*.

The whole population of the county is a little more than 600,000, or not much greater than that of Liverpool and Birkenhead.

III. **Cornwall** forms the extreme south-western part of England, thrust out between the Atlantic Ocean and the English Channel, and ending in two rocky points, the Lizard and the Land's End. It has thus but one land boundary, that on the north-east toward Devonshire, from which it is separated by the

Tamar. The county is little more than half the size of Devon, having an area of 1,365 square miles. It consists of high and barren moorlands, with steep rounded hills of granite, and valleys of boggy soil. The central ridge of heights has a rapid slope on either side, furrowed by the valleys of many small rivers, such as the *Camel* on the north side, and the *Fal* on the south, whose estuaries cut deep into the coast. The moorlands reach their greatest height in the north-east, where *Brown Willy* rises to 1,368 feet; from thence they gradually fall to 600 and 800 feet near Mount's Bay.

In the wild and thinly-peopled moorlands between the Tamar, and the Fal, the only town of any size is *Bodmin*, the county town, in the valley of the Camel. Throughout the interior of the county, indeed, the villages consist merely of a few people gathered round the shaft of a mine, and fully nine-tenths of the whole population of the county has been drawn down to a little space in the southern end of the peninsula, where the rocks are richer in varied mineral wealth than in any place throughout all the rest of England. Here are mines of copper and lead and silver, and rarer than all these, of tin, which have long made the wealth of Cornwall. *Truro, Camborne,* and *Redruth* are the chief towns of this mining district. Along the coast, and forming the outlets of the mining trade, are the ports of *Falmouth,* on Falmouth Harbour; *Penzance,* on Mount's Bay; and *St. Ives,* on St. Ives Bay. The population of the whole county amounts to 362,343 persons.

County.	County Town.	Area in Sq. Miles.	Population.
(Northern Counties.)			
Northumberland .	Newcastle .	1,952	386,646
Durham	Durham . .	973	685,089
Cumberland . .	Carlisle . .	1,565	220,253
Westmoreland . .	Appleby . .	758	65,010
(Counties of the Mersey and Ribble basin.)			
Lancashire . . .	Lancaster .	1,905	2,819,495
Cheshire . . .	Chester . .	1,105	561,201
(Counties of the Severn basin.)			
Shropshire . . .	Shrewsbury .	1,291	248,111
Worcestershire .	Worcester .	738	338,837
Gloucestershire .	Gloucester .	1,2;8	534,640
Warwickshire . .	Warwick . .	881	634,189
Herefordshire . .	Hereford . .	836	125,370
Monmouthshire .	Monmouth .	576	195,448
(Counties of the Humber basin.)			
Derbyshire . . .	Derby . . .	1,029	379,394
Staffordshire . .	Stafford . .	1,138	858,326
Leicestershire . .	Leicester . .	80;	269,311
Nottinghamshire .	Nottingham .	822	319,758
Yorkshire . . .	York . . .	6,067	436,355
(Counties of the Wash.)			
Lincolnshire . .	Lincoln . .	2,776	436,599
Rutland	Oakham . .	150	22,073
Northamptonshire	Northampton	985	243,891
Huntingdonshire .	Huntingdon .	361	63,708
Bedfordshire . .	Bedford . .	462	146,257
Cambridgeshire .	Cambridge .	818	186,906

County.	County Town.	Area in Sq. Miles.	Population.
(East Anglian Counties.)			
Norfolk	Norwich . .	2,116	438,656
Suffolk	Ipswich . .	1,481	348,869
(Counties of the Thames basin.)			
Essex	Chelmsford .	1,657	466,436
Hertfordshire . .	Hertford . .	611	192,226
Middlesex . . .	Brentford. .	· 281	2,539,765
Buckinghamshire.	Aylesbury .	730	175,879
Oxfordshire . .	Oxford . .	739	177,975
Berkshire . . .	Reading . .	705	196,475
Surrey	Guildford. .	748	1,091,635
Kent	Maidstone .	1,627	848,294
(Counties of the Southern Water-parting.)			
Sussex	Chichester .	1,460	417,456
Hampshire . . .	Winchester .	1,672	544,684
Wiltshire . . .	Salisbury . ·.	1,352	257,177
Dorsetshire . .	Dorchester .	987	195,537
(The South-western Counties.)			
Somersetshire . .	Taunton . .	1,636	463,483
Devonshire . .	Exeter . .	2,589	601,374
Cornwall . . .	Bodmin . .	1,365	362,343

WALES.

CHAPTER I.

GENERAL VIEW OF WALES.

GENERAL FEATURES.—Wales, the peninsula which is thrown out from the western side of Southern Britain towards Ireland, forms a region which is distinguished from England at once by its geographical position, by the age of its rocks, by the height of its mountains and grandeur of its scenery, and by the origin of its people.

Geographically, the country lies apart from the rest of Southern Britain (*see* Frontispiece). To the north it is bathed by the Irish Sea, to the south by the Bristol Channel, while to eastward it was once parted from England by that deep and remarkable depression of the ground which strikes southward from the estuary of the Dee to the estuary of the Severn. The political changes which have added a part of the original country of the Welsh people to England have, however, pushed the eastern boundary of Wales backwards, so that it

now forms a straight line which runs due south from
the mouth of the Dee, cuts across the upper valleys
of the Severn and the Wye, passes over the eastern
extremity of the Black Mountains, and terminates
in Cardiff Harbour. To the west of this line rise the
greater mountains of the Welsh group ; to the east
of it lie tracts of level country, or outlying ranges of
lower hills which fall by degrees to the central plain
of England. The boundary of Wales to the east thus
agrees, roughly speaking, with the change from the
mountains to the plain.

The country which lies within these limits is, in
comparison with England, of very small extent, for it
includes but 7,400 square miles, or an area only equal
to one-twelfth part of Britain. Consisting as it does
practically of two short mountainous peninsulas, it differs from England no less in structure than in scenery;
for the Silurian and Devonian rocks, the masses of
slate and limestone and porphyry, of which it is built
up, represent the remains of an ancient Britain which
rose above the ocean at a time when the England
that now exists was being slowly formed in its depths.

But while the mountains of Wales thus contrast
with the lowlands of England, the two countries have
one feature of their physical structure in common. In
Wales as in England, the chief mountain heights lie
near the western coast ; and the line of water-parting
which they form strikes down the country from north
to south, and throws the great bulk of its running
waters to the eastward (see p. 77). The relative position of mountains and river-valleys is therefore the

same in both Wales and England, though the relative extent of the hills and valleys is so different.

As we have already seen, the mountainous structure of Wales has had very marked effects on the history and character of the country. (1) In the rocky fastnesses of its hills the ancient race of the Celtic people has been preserved, a people who still retain their old customs and language. (2) Again, the disposition of its mountains in great measure determines the climate of the country as well as its excessive rainfall. (3) By the physical character of the ground, too, the agricultural industry of Wales is strictly limited. It is only in the valleys near the sea that the ground is fit for ploughing; tillage is therefore very scarce, and the chief agricultural wealth of Wales lies in its mountain pastures. On the other hand, the inexhaustible stores of slate, and above all, of coal, which enrich its barren hills, give rise to a very considerable trading and manufacturing industry. (4) Lastly, the structure of Wales has greatly hindered the increase of its population. The whole country is about one-seventh the size of England, but owing to its mountainous character it has only 1,200,000 inhabitants, or one-eighteenth part of the population of England. It is more than 1,000 square miles greater in extent than Yorkshire, but it only contains half as many people as that county, or one-third of the inhabitants of London.

THE MOUNTAIN MASSES of which Wales is made up are broken into four distinct ranges, which all lie athwart the line of water-parting. These are (1) the

Snowdon range ; (2) the *Berwyn Mountains ;* (3) the
Plinlimmon range ; (4) the *Black Mountains* (see p.
46). The first three ranges have the same general
direction, their strike being from north-east to south-
west ; they are alike also in structure, being mainly
composed of slate rocks of the Silurian age, broken by
masses of trap and porphyry of volcanic origin. The
fourth range, on the other hand, differs from the rest
not only in having a direction due east and west,
but also in its structure ; for the limestone rocks of
which it consists belong to the somewhat later Devonian age, and are overlaid on the southern slopes of
the Black Mountains by vast coal-measures which
exceed in extent and in depth any of those in England.
We shall best gain a general idea of the true character
of Wales by examining these four mountain chains in
detail.

(1.) The **Snowdon Range** is situated in the north-
west corner of Wales, where it stretches in a long line of
steep heights along the Menai Straits and Carnarvon
Bay, and forms the northern boundary of Cardigan
Bay. On the east it is sharply cut off from the other
Welsh mountains by the deep valley of the river
Conway.

The central mountain of this range is *Snowdon,*
whose five peaks of nearly equal height rise 3,590
feet above the sea-level, and from their position and
elevation determine the river system of the surrounding district. To the north the lower spurs of the
mountain form the *Pass of Llanberis,* beyond which
lies a group of heights almost as lofty as Snowdon

itself, the *Glyders, Carnedd Davydd,* and *Carnedd Llewellyn,* the highest point of a chain which strikes northwards to the sea, and terminates in the bold cliffs of *Great Orme's Head.* All these summits are higher than any in England ; but they form the only Welsh mountains which reach an elevation of 3,000 feet above the sea. On the eastern side of Snowdon towards the Conway the greatest height is that of *Moel Siabod,* 2,865 feet; while to the south-west a line of hilly country is thrown out which falls rapidly to the sea in the peninsula of *Carnarvon,* and ends in the headland of *Braich-y-Pwll.*

The wild and rugged scenery of this district surpasses in grandeur any other mountain scenery in Britain south of the Tweed. The country is bare and uncultivated, for amid its rocks of slate and granite there is little room for tillage ; it is traversed by few roads, and very thinly inhabited, its villages being gathered round the slate quarries or the lead and copper mines which form their only wealth. The chief resources of northern Wales lie, in fact, in the rocks of which its mountains are composed. The inexhaustible stores of slate contained in the district round Snowdon are worked in quarries of vast extent, such as those of Penrhyn, which give employment to over 3,000 workmen, and yield every year 70,000 tons of slate, large quantities of which are exported to Norway, and even to America, through a special port named Port Penrhyn, which is connected with the quarries by a railway and is wholly engaged in this trade.

(2.) To the south and east of the Snowdon range

lies a second tract of mountainous country of very
irregular character, whose south-eastern boundary is
formed by the chain of the **Berwyn Mountains**,
a long line of hills which stretches from the valley of
the Dovey in the south-west as far as the vale of Llangollen or the valley of the Dee in the north-east. A
second belt of high ground called the *Flint Hills*
starts from the further side of this valley and runs due
north to the Irish Sea between the estuary of the Dee
and the valley of the Clwyd. The space inclosed between the Berwyn Mountains, the Flint Hills, and the
Snowdon range is occupied by an irregular mass of
mountainous country that touches the Irish Sea on the
north and Cardigan Bay on the south-west, and divides
itself into two parts which differ in structure as in
scenery.

(*a*.) The northernmost district, which lies on the
Irish Sea, consists of a tract of tumbled ground,
known as the *Denbigh Hills*, whose limestone rocks
never rise to any considerable height, but extend in
irregular hills over the region which lies between the
Conway and the *Clwyd*, two rivers whose channels
form parallel valleys that open into the Irish Sea.
Almost the whole agricultural industry of the district
is gathered into these two fertile river-valleys ; while
the valley of the *Dee*, which lies to the south and
east, is distinguished for the mineral resources which
it possesses in the small coal-field of *Ruabon*, and
that which extends along its estuary in Flintshire.

(*b*.) The south-western district, on the other hand,
which lies between the Snowdon and the Berwyn

Mountains, and stretches along the shores of Cardigan Bay to the valley of the *Dovey*, consists of a mass of slate mountains, very lofty and wild, and broken as in Snowdon by trap rocks of volcanic origin. The centre of this district is marked by the mountains of *Arrenig* and *Arran Mowddy*, in which last summit the slate rocks attain a height of nearly 3,000 feet; and the whole surrounding region abounds in quarries, such as those of Ffestiniog, which are equal in extent to any in the Snowdon range. To the south-west a deep depression severs the bulk of these slate hills from the great mass of *Cader Idris*, where the volcanic rocks reach their greatest height of 2,900 feet. As this same depression strikes in a north-easterly direction under the heights of the Berwyn chain, it forms the upper valley of the river *Dee*, and is marked by the only large sheet of water in Wales, *Bala Lake*, which is traversed and drained by the Dee.

(3.) A third belt of high ground is formed by a long chain of hills and moorlands which strike southward from the mountain of Plinlimmon, and sweep in a curve round the shores of Cardigan Bay to end in the south-west in the peninsula of Pembroke. The chief mountain of this chain, Plinlimmon, forms an almost isolated mass of 2,470 feet in height, which occupies a middle position between the heights of northern and southern Wales, and forms the connecting link between them. Plinlimmon may be looked on as the great centre, which, by its outlying hill ranges, and by its rivers, determines the surrounding geography of the district in which it lies. Situated

due south of Arrenig and Arran Mowddy, it marks
the most important point in the line of water-parting,
the district in which most of the larger streams of
Wales take their rise.   On its eastern side it throws
down the great rivers of the *Severn* and the *Wye*, and
on the west the little stream of the *Ystwith;* while
a short distance to the south are the head-waters of
the *Towy* and the *Teifi.*   The channels of these two
last rivers form parallel valleys, which define and enclose between them a long belt of hill and moorland
that trends to the south-west, and terminates on the
shore of St. George's Channel in the peninsula of
Pembroke.   These moorlands are indeed of incon-
siderable height, and the lower grounds and valleys
of the district, where the plough can be used, form
the best agricultural land in Wales.   On the other
hand, the country which lies to eastward of Plinlimmon as far as the borders of Wales is utterly wild
and barren.   Its vast tracts of moorland, broken
by hills and mountain ridges of limestone, are wholly
worthless for tillage, and remain the most thinly-
peopled regions of southern Britain.

(4.) If we now strike due southward from Plinlimmon, following the course of the Towy, we find
ourselves at the western extremity of the **Black
Mountains**, the chief mountain chain of southern
Wales.   This is a range of great height which runs
due east and west in a direction differing from that
of the other Welsh ranges ; its northern side is sharply
defined by the valley of the *Usk*, from which the mountains rise in a steep and abrupt wall of limestone cliffs,

and reach in *Brecknock Beacon* a height of 2,900 feet. To east and west the range is limited by the valleys of the Usk and the Towy; while to the southward its heights fall by a long and gradual descent to the Bristol Channel, having their slopes furrowed by a multitude of little streams, such as the *Tawe*, the *Neath* and the *Taff*, with others, whose outlets to the sea form excellent harbours.

This mountain chain, more regular in form than those of the north, and unbroken by masses of volcanic rocks such as appear in the other ranges, is far richer than they are in mineral wealth. While the whole of northern Wales, whose chief resources lie in its slate-quarries, possesses but a small tract of coal in the Dee valley, the southern slopes of the Black Mountains are covered by a coal-bed whose extent, 900 square miles, exceeds that of any other in southern Britain, and which reaches a depth of 10,000 feet. Here, therefore, as in the mining districts of the Pennine chain, the solitude of the moorlands is exchanged for the busy life of a manufacturing district, with its mines, its blast-furnaces, its iron-works, and its copper-foundries. The little river valleys are thickly studded with mining towns, while the harbours at their estuaries provide outlets for the trade in coal and iron which has made this district the most wealthy and important in Wales, so that one-third of the whole population of the country is gathered into it.

The industry of this coal-field, indeed, is not confined to the working of the minerals which it

produces; for by the great advantages of its geographical position, and the quality of its mineral deposits, it
has become also the centre of a busy foreign trade,
which is carried on mainly through two seaports
which lie to eastward and westward of the coal-measures. The port of *Cardiff* is engaged in the export
of coal and iron, and ranks as the tenth or twelfth in
importance among European harbours. The port of
Swansea, on the other hand, has become the first
town in the world for the smelting of copper and
other metals, and from the neighbouring peninsula
of Cornwall, from France, from North and South
America, and from Australia, large quantities of
metal are carried to its foundries.

THE RIVER SYSTEM.—In this sketch of the mountain masses of Wales, we have been led to notice the
main river valleys by which the strike of the hill
ranges is defined, and which break the long reaches
of moorland that form so large a part of the country.
We must now, however, glance briefly at the river
system of Wales taken as a whole.

The division between the running waters of the
country is indicated by a line which branches off from
the main waterparting of England as it passes through
the plain of Cheshire (see p. 101). From this point
the Welsh waterparting strikes to west and south
over the chain of the Berwyn Mountains, across the
summit of Plinlimmon, and along the crest of the
heights which form the peninsula of Pembroke. It
thus divides the country into two parts, the *watershed
of the Irish Sea*, lying to the north and west, and the

watershed of the Bristol Channel, which lies to the south and east. These divisions are very unequal in the extent and importance of their river system, for as we have seen, the high grounds of Wales lie like those of England near its western shores, and the bulk of its running waters are thrown down to the east and south. In fact the only river of any consequence in the basin of the Irish Sea is the Dee, which flows to the eastward for the greater part of its course, and only turns northward as it nears its estuary. The basin of the Bristol Channel, on the other hand, includes the upper valleys of all those great Welsh rivers which merge their waters in the estuary of the Severn (see p. 46).

As the more important among the Welsh rivers have been described in Chap. VII., it will be only necessary here to sum up shortly the groups into which they are naturally thrown by the direction of their channels and by their openings into the sea.

I. **The Rivers of the Irish Sea** form two groups. (*a*) Those which fall into the sea north-ward, as the *Dee*, the *Clwyd*, and the *Conway*. (*b*) Those which fall into the sea westward, as the *Dovey*, the *Ystwith*, and the *Teifi*. All these rivers, save the Dee, are mere mountain streams, and quite useless for navigation or harbourage (see p. 46).

II. **The Rivers of the Bristol Channel** in like manner form two groups. (*a.*) Those which merge in the estuary of the Severn, the *Severn*, the *Wye*, and the *Usk;* these are all rivers of import-ance, but it is merely along their upper valleys, where the streams are but mountain torrents, that these rivers are included in Wales. (*b.*) Those streams which fall southward into the Bristol Channel, such as the *Taff*, the *Neath*, the *Tawe*, and the *Towy*, are severed from the last group by the range of the Black Mountains. The three first streams, which are thrown down the southern slopes of that mountain range, are short and unimportant save for their harbours. The Towy on the other hand, rising among the hills of the Plinlimmon range, has a longer course ; and its deep valley forms a clearly marked geographical boundary between the slates and porphyry of the hills to westward of it, and the limestone rocks and coal measures of the Black Mountains. (See p. 96.)

The river system of Wales has few points of resemblance with that of England, whether we consider the size of its streams, the extent of their basins, or the length of their courses. Nevertheless they play a very important part in the political geography of the country. They indicate the course which must be followed by the roads and railways which traverse the land, and through them, therefore, all communication between different parts of the country passes: while it is only along their banks that agriculture can be pursued, and that towns and villages can spring up. Hence we find, as we have already seen in England, that it is the river-basins of Wales which have determined its political divisions, and that we shall best group the shires into which it is broken up by following carefully the lie of the two watersheds which form the natural divisions of the land.

CHAPTER II.

WALES is broken up for political purposes into *twelve* distinct shires or counties, of which six are comprised in the watershed of the Irish Sea, while six more form the watershed of the Bristol Channel. From the form of the country, a short peninsula washed on three sides by the sea, it follows that nearly all these shires are situated on the coast, and in fact there are but three among them which lie inland.

[A.]—THE COUNTIES OF THE IRISH SEA extend over an area of about 3,000 square miles, and comprise the ranges of the Snowdon and the Berwyn Mountains with the hilly country that lies between them, and part of the Plinlimmon range. They thus include the chief mountains of Wales, and form a wild and uncultivated district which contains but one-third of the Welsh people, or 458,614 inhabitants, a population somewhat less than that of Manchester. All these shires lie directly on the sea-coast.

I. The **Island of Anglesea,** which forms a
distinct county, is separated from the mainland by the

THE WELSH COUNTIES.—(Towns and Rivers.)

narrow Menai Strait, and is the only Welsh county
whose surface is generally low and flat. It has an

area of 300 square miles, and a population of about
50,000 people. Bathed in soft sea-winds, the fertility
of its soil is increased by the mildness of its climate ;
and it possesses a little mineral wealth in a small
coal-field, as well as some rich veins of copper. Its towns
are few ; the trade with England is carried on through
Beaumaris on the Menai Strait, and *Almwch* on the
northern coast is situated near some small copper-
mines.   A second little island, *Holy Island*, lies close
to its western shore, and contains the port of *Holyhead*,
with a town of the same name which forms the point
of departure of travellers from England to Ireland.

II. The county of **Carnarvon**, which closely
adjoins Anglesea, lies within sharply defined limits,
since it consists exactly of the rugged mass of the
Snowdon range and the sharp spur of rocks which
runs southward from it to the sea.   The shire thus
extends from Great Orme's Head to Braich-y-Pwll.
Its form is that of a triangle, with an area of 579
square miles : on the east it is parted from Denbighshire by the river Conway; on the south it is
bounded by Merioneth.   The mountain masses which
constitute the county form a bleak and inhospitable
region which contains but a few villages situated
beside the great slate-quarries, such as *Penrhyn*.   The
larger towns lie of necessity in the low ground along
the shore, where *Carnarvon, Bangor,* and *Conway,* on
the north-western coast, occupy the sites of three
great castles built by the English 600 years ago to
hold their conquests in Wales.   Carnarvon, the largest
town of northern Wales, with 10,000 inhabitants,

guards the southern opening of the Menai Strait: Bangor, a cathedral city, commands the junction of two great routes, that which leads westward to Ireland, and that which turns southward through Wales: Conway holds the estuary of the river Conway and the northern opening of the Menai Strait. Watering-places are also scattered along the seaside, as *Llandudno*, under the rocky promontory of the Great Orme's Head. The whole population of the shire amounts to about 106,000 inhabitants.

III. To eastward of Carnarvon lies the shire of **Denbigh**, which stretches in a slanting direction from the Irish Sea on the north to the English border on the south-east. Its area is 600 square miles, but its form is very irregular, being so narrowed at the centre by the shires of Flint and Merioneth on either side of it as to have an appearance not unlike that of an hour-glass. The county thus practically consists of two parts. (*a.*) Its northern half lies between Carnarvon and Flintshire, being severed from them by the valleys of the *Conway* and the *Clwyd*. It is chiefly composed of broken and tumbled ground, and low hills which fall gradually to the sea. The little agricultural towns, *Ruthin* and *Denbigh*, lying in the valley of the Clwyd, and situated at a distance from any great lines of trade, are without importance. (*b.*) The south-eastern part of the shire, on the other hand, forms part of the valley of the *Dee*, which, as it enters Denbighshire, bends eastward round the extremity of the Berwyn Mountains and forms the *Vale of Llangollen*. When it passes out of this vale the river enters on a small tract of

coal-measures which gives rise to two mining towns,
Ruabon and *Wrexham*, whose position near the
English border gives them a considerable trade with
Liverpool and the manufacturing towns of the west.
The population of the shire, 105,100 persons, is about
equal to that of Carnarvon.

IV. If we follow the course of the Dee as it flows
out of Denbighshire to the northward, we are led
directly into **Flintshire,** the smallest of the Welsh
counties, being merely a narrow strip of land with
an area of 289 square miles, which borders the estuary
of the Dee on its western side, and is itself shut in to
south and west by the shire of Denbigh : an outlying
fragment of the county is imbedded in Shropshire to
the south-east.   If Flintshire, however, is the least of
the counties of Wales, it is also for its size one of the
richest, the busiest, and the most thickly peopled.
The coal-measures, which here extend along the
valley of the Dee from the Flint Hills to the river
bank, with their mines of iron, zinc, and lead, give
rise to a good deal of manufacturing industry, while
their neighbourhood to the sea and to some of the
great manufacturing districts of England affords the
needful facilities for commerce.   The town of *Mold,*
in a lateral valley of the Dee, has mines of lead and
coal; *Holywell*, near the Dee itself, lies in the midst
of the coal district, and exports its minerals to Liverpool through the little port of *Flint*, which gives its
name to the county.   The only other towns in the
shire lie in the lower valley of the Clwyd : *St. Asaph*
is now a mere village, though it still remains an

episcopal city ; *Rhyl* is a fashionable watering-place, situated at the mouth of the river. The industrial resources of Flintshire enables it to maintain a population of about 76,300.

V. These counties all border on the Irish Sea, and are watered by the rivers which flow northward to that sea. The two remaining shires of this group lie on the shores of Cardigan Bay, and are drained by those streams which are thrown down the western slopes of the waterparting.

Due south of Carnarvon, and extending from that shire to the Berwyn Mountains and the border of Montgomery, lies the county of **Merioneth**. Its form is like Carnarvon that of a triangle, with an area of about the same size, 600 square miles, but turned in the opposite direction. The narrow eastern extremity, thrust as it is into the heart of Denbighshire, consists practically of the upper valley of the *Dee*, which with the hills on either side of it forms a wild and solitary district without a single town. The broad western end rests on Cardigan Bay between the shires of Carnarvon and Cardigan, and is bounded on the south by the valley of the *Dovey*. From *Bala Lake* to the sea-shore it is composed of a mass of rugged and barren mountains, amidst which tower the great summits of *Arran Mowddy* and *Cader Idris*. Its bleak heights are uninhabited, save for a few villages gathered round its slate quarries, as at *Ffestiniog;* and for the very small towns which lie on or near the sea-shore—*Harlech*, the county town, *Barmouth* and *Dolgelly*. In fact the shire of Merioneth, with its

46,600 inhabitants, is one of the most thinly peopled counties in Wales.

VI. To the south of Merioneth lies the last county of this group, **Cardiganshire**; it is almost isolated from all the other shires belonging to the watershed of the Irish Sea, being half encircled by the counties of the south-western watershed, those of Montgomery, Radnor, Brecknock, Carnarvon, and Pembroke. The mountain of Plinlimmon, which rises just within its northern limits, sends out a long spur of heights which run to the south-west, and the shire of Cardigan lies in the form of a great crescent upon the western slope of this range, from the valley of the Dovey in the north to that of the Teifi in the south. The little stream of the Ystwith flows midway between the two. The hill-sides have their mines of copper, and lead, and zinc, and the river valleys afford the possibility of agriculture, so that there is a somewhat larger population than that of Merioneth. Still with an area of 700 square miles the county has but 73,441 inhabitants. It contains, in fact, only two towns of any note, and these lie on the sea-shore. The county town, *Cardigan*, at the mouth of the Teifi, is little more than a fishing village; the town of *Aberystwith*, at the mouth or " aber " of the Ystwith, is more populous.

[B.]—THE COUNTIES WHICH DRAIN THEIR WATERS INTO THE BRISTOL CHANNEL cover an area of over 4,300 square miles, and comprise the greater part of the Plinlimmon range, with the Black Mountains. While they contain fewer lofty mountain summits than the

shires of the north and west, they include yet larger
tracts of desolate and unpeopled moorlands broken
by rough hills.   At the same time this group of
counties possesses sources of wealth not open to the
northern group in its coal-measures, and in its broad
river valleys where it is possible to plough and till the
land.   It is thus the most industrious and populous
part of Wales, containing two-thirds of the people of
the country, or a population of over 758,000 people ;
less than that of the single English county of Kent.

In this group are included the three Welsh shires
which lie wholly inland : these three shires, Montgomery, Radnor, and Brecknock, border on English
soil, and form the upper basins of the three rivers
which unite their waters in the estuary of the Severn.

I. The northernmost county of this group, Montgomeryshire, forms a rough circle almost wholly
surrounded by hills.   To the north and north-west
the Berwyn Mountains shut it in from the shires of
Denbigh and Merioneth ; to the south-west and south,
Plinlimmon and its outlying moorlands form the
borders of Cardigan and Radnor; while to the east,
where it adjoins England, a line of heights parts it
from Shropshire.   The shire thus enclosed on every
side is of considerable size, having an area of 755
square miles, and consists in fact of the upper basin
of the *Severn*, that is, the whole of that part of its
basin which lies within Wales.   The Severn rises
on the hills to the south of the shire, near Plinlimmon, while its tributary the *Virnwy* rises on the
Berwyn Mountains of its northern boundary ; and both

rivers unite their waters as the Severn escapes by
the only opening in the encircling heights, a pass
under the *Breidden Hills*, which leads into English
ground. The bleak moorlands of slate and limestone which compose so great a part of Montgomery
are thus bröken by two river valleys which traverse
the shire from the north and from the south, and
in the shelter of these valleys has sprung up the
only industry of this district, the manufacture of
Welsh flannel. In the larger valley of the Severn
especially, a few towns have grown up by the river
side; *Llanidloes* and *Newtown, Montgomery* the
county town, and *Welshpool* near the English border,
somewhat more prosperous than the rest. The
reaches of moorland and heather which extend between the river valleys and stretch up the hill-sides
of the western border are wholly deserted save for
a hamlet here and there, and for the mountain
cattle feeding on their slopes; and the shire, with
its extent of country equal to nearly three times
that of Flintshire, contains but 67,623 inhabitants,
and forms one of the most thinly peopled districts
in Wales.

II. We have traced the waters of the Severn as
they traverse the county of Montgomery to the north-
east; but close to the sources of the Severn lie the
headwaters of the Wye, and this river as it turns to
the south-east forms by its upper basin the shire of
Radnor.

Lying on the eastern slopes of the Plinlimmon
range, as Cardiganshire lies upon the western, the

county of Radnor extends eastward to the English border between Montgomery and Brecknock. The Teme, an affluent of the Severn, passes along its north-eastern boundary, parting it from Shropshire : but the great bulk of the running waters of the shire are poured into the Wye, as this river winds along its southern border on its way to Herefordshire in the east. The river basin which thus constitutes Radnorshire has an area of more than 400 square miles, and forms one continuous tract of high, bleak moorland, broken by barren mountains, its chief heights lying in the district called *Radnor Forest*. The desolate monotony of this wild region is everywhere unbroken by the stir of human life ; for but a few villages lie along the eastern border, where they find shelter in the river valleys : these are *New Radnor* on the Wye, *Presteign* on its tributary the Lug, and *Knighton* on the Teme. The population, 25,430, is smaller than in any other Welsh county.

III. From the shire of Radnor we pass southward to that of **Brecknock**, which forms the third inland county of Wales, and practically consists of the upper basin of the Usk, the third great river of the Severn estuary. This shire is one of the largest in Wales, having an area of 719 square miles. Right through its centre strikes the deep valley formed by the Usk, as the river rises on the western border of the county and passes eastward to Monmouthshire, taking up on its way the mountain streams thrown down to it from north and south. On its banks lies the little county town, *Brecon*, which marks the central point of the

shire ; while to north and south the slopes of the river
basin form two districts that differ considerably in
structure. (*a*) To the northward the rocks of Devonian
limestone of which the valley of the Usk consists
pass into the slate rocks of the Plinlimmon range,
and form a district with the same bleak moorland
character as that of Radnor, which is only parted
from it by the Wye. There are but two towns in
this region ; these are *Builth* and *Hay*, situated on the
border of the county by the Wye. (*b*) ·To the south
of the Usk valley rises the steep range of the Black
Mountains, whose highest point, Brecknock Beacon, is
lifted 2,860 feet above the sea. This range, built up
in great part by the mountain limestone and millstone
grit which form the Pennine chain, extends westward
into the shire of Carmarthen, and falls southward to
that of Glamorgan. It contains no towns ; in fact
the wild and inhospitable region formed by the bleak
moorlands and mountain masses which constitute
Brecknockshire maintains a very scanty population ;
for the whole shire has but 60,000 inhabitants.

IV. In describing the county of Brecknock we
have traced the course of the chief mountain range of
Southern Wales, the Black Mountains, and Brecknock
Beacon. The slopes of this range as they fall to
the Bristol Channel on the south are occupied by the
shire of **Glamorgan,** which thus forms the southernmost county of Wales, and from the border of
Monmouth on the east to that of Carmarthen on
the west, comprises the basins of all the streams
which are thrown down from the crest of .the Black

Mountains, such as the Tawe, the Neath, and the Taff.
It is the second among Welsh counties in point of
size, having an area of 856 square miles; and it is in-
dustrially the most important part of Wales, since its
limits almost exactly coincide with those of the vast
coal-measures which cover the hill-sides for a distance
of 900 square miles. Glamorganshire forms, therefore,
one of the chief mining districts in Great Britain.
The abundance of its coal and iron, the complete
system of running waters which it possesses, its geographical position, and the excellent harbours along
its coast, combine to secure it a foremost place both
in manufacturing industries and in foreign trade.
The great centre of its mining industry lies to the
eastward in the valley of the Taff. High up in this
valley, on the northern border of the shire, the coal-
mines have given rise to the most populous town of
Wales, *Merthyr Tydvil*, with 52,000 inhabitants and
extensive iron works; not far off lies a second mining
town, *Aberdare*, containing 36,000 people; and to
eastward the great furnaces of *Dowlais*, where 20,000
workmen are sometimes employed. From this point
southward the Taff valley presents a long succession
of manufactories and mining villages, till the river
empties itself into the Bristol Channel, and forms the
great trading port of this district, Cardiff Harbour, on
whose shore lies the town of *Cardiff* with 40,000 inhabitants. Close by this important modern town lies the very
ancient episcopal city of *Llandaff*, now a mere village,
though it still gives its name to a bishopric; it is
situated in the midst of a fertile district of low

land known as the *Vale of Glamorgan*, which extends
westward from the lower valley of the Taff as far as
Swansea Bay. The other towns of note lie on the
shores of this Bay; *Neath*, on the river of the same
name, is engaged in exporting coal; *Swansea*, at the
mouth of the Tawe, is the great centre of the iron
manufacture and of the trade with Europe, America,
and Australia in metals, especially copper, for smelting: it is the largest Welsh seaport and contains
52,000 inhabitants.

Glamorganshire is by far the most thickly peopled
county in Wales; with its population of 397,859 it
contains little less than one-third of the people of the
country, and has almost as many inhabitants as are to
be found in all the counties of the Irish Sea if we
except Anglesea. There are but four towns in Wales
whose population exceeds 15,000, and these four
towns lie in Glamorganshire.

V. The chain of the Black Mountains passes from
the borders of Glamorgan and Brecknock westward
into the shire of Carmarthen, where it is cut short
by the valley of the river *Towy*. The basin of this
river constitutes in fact the bulk of Carmarthenshire.
As the stream crosses the county from north to south-
west, it severs between the limestone rocks of the
Black Mountains to the eastward and the slate moorlands of the Plinlimmon range on the west, moorlands
which extend across its border on this side from the
shires of Cardigan and Pembroke. From these heights
a number of little streams are thrown down on either
side to the central valley of the Towy, and in this

valley lie almost all the towns of the shire, most of them very small. *Llandovery* is situated high up in the valley; *Llandeilo*, lower down, marks the site of a castle of the old princes of South Wales; *Carmarthen*, the county town, is a port of considerable size lying at the mouth of the river. The only town situated outside the Towy valley is the port of *Llanelly* on the Bury estuary, which lies just within the limits of the coal-measures, and has a considerable trade. The county of Carmarthen is the largest in Wales, having an area of 950 square miles : its population is 115,710.

VI. The moorlands of the Plinlimmon range which occupy the north and west of Carmarthen pass out of that county into **Pembrokeshire** on the south-west, where they form the peninsula in which Wales terminates in this direction. This shire has therefore but one land boundary, that on the north-east towards Cardigan and Carmarthen; on all other sides it is surrounded by sea, and has its shores worn into deep bays and creeks such as *St. Bride's Bay*, and *Milford Haven*, a harbour large enough to contain the whole of the British fleet. The scenery of Pembrokeshire is not very rugged; its slate rocks are broken by masses of trap and porphyry, as in the peninsula of Carnarvon; but throughout the whole shire, as the hills near the sea, they sink so low that large tracts of land can be cultivated. The chief towns, however, all lie round the inlet of Milford Haven, to which population is drawn by its shipping trade and its yards for ship-building. The county town,

Haverfordwest, is situated at the head of the haven ;
Pembroke, on its southern shore, is the largest town
in South Wales outside the manufacturing districts ;
Milford, on the opposite side, is but a small town.
The episcopal city of *St. David's* at the extremity of
the peninsula by St. Bride's Bay is little more than a
village ; *Tenby* is a picturesque watering-place which
rests on the shore of Carmarthen Bay.

With an area of over 600 square miles, Pembrokeshire maintains a population of about 92,000 people.

Counties of the Irish Sea Basin.

Counties.	County Towns.	Area in sq. miles.	Populat on.
Anglesea.	Beaumaris.	302	51,040
Carnarvonshire.	Carnarvon.	579	106,121
Denbighshire.	Denbigh.	603	105,102
Flintshire.	Mold.	289	76,312
Merionethshire.	Harlech.	602	46,598
Cardiganshire.	Cardigan.	693	73,441
		3,068	458,614

Counties of the Bristol Channel Basin.

Montgomeryshire.	Montgomery.	755	67,623
Radnorshire.	Presteign.	425	25,430
Brecknockshire.	Brecon.	719	59,901
Glamorganshire.	Cardiff.	856	397,859
Carmarthenshire.	Carmarthen.	947	115,710
Pembrokeshire.	Haverfordwest.	628	91,998
		4,330	758,521
	TOTAL ...	7,398	1,217,135

The **ISLE OF MAN**, though forming part of the English dominions, retains its own government and judges. Its inhabitants, the Manx people, belong to the Keltic race.

This island is situated in the Irish Sea, midway between England, Scotland, and Ireland. In point of size it is smaller than Anglesea, having an area of but 220 square miles; it is about thirty miles long, and has an average breadth of ten miles. Its scenery is of a mountainous character; in fact the island practically consists of a short range of high hills which strikes from north-east to south-west, and rises in its chief summit, that of *Snaefell*, more than 2,000 feet above the sea-level. A fringe of low lands borders this range of hills, and contains the little towns of the island, which are all very small save the capital, *Douglas*, a town of 14,000 inhabitants.

The Isle of Man is very rich in minerals, containing veins of lead, copper, silver, and iron; besides these it has valuable quarries for slate and building-stone. With these mining industries it is enabled to maintain 54,042 inhabitants, a population which slightly exceeds that of Anglesea.

SCOTLAND

PHYSICAL

0   0   20   30   40   50
English Miles

SCOTLAND.

CHAPTER I.

Boundaries.—In the northern part of the island
of Great Britain lies SCOTLAND, a country which is
distinguished from England both by position and by
striking differences of physical structure.

To the north and west it is washed by the *Atlantic
Ocean*, and to the east by the *North Sea ;* while to
the south it is cut off from England by an arm of the
western sea called the *Solway Firth*, which penetrates
far into the land, and by a chain of mountains, the
Cheviot Hills, which are thrown across the island from
the head of the Firth to the river *Tweed*. This wall of
heights, preventing all entrance into Scotland from the
south save along the low ground by the coast on either
side, forms a natural barrier that divides Great Britain
into two parts, and which long constituted the boundary of two distinct kingdoms.

Area.—The country which lies within these limits has an area of 26,000 square miles,—and is therefore only half the size of England. But the great number of islands which cluster round its shores have, taken together, an extent of 4,000 square miles, and by the addition of these islands the whole area of the Scotch kingdom is increased to 30,000 square miles, or nearly the size of Ireland. Its most remarkable feature is the disproportion between its length and breadth, which is nearly twice as great as that of England; for while Scotland is 270 miles long it has nowhere a breadth of more than 160 miles, and measures little more than 30 miles across its narrowest part. It thus presents the appearance of a long and narrow barrier of rock thrown up between the oceans which dash against it from east and west—a barrier which has in places been almost cut in two by the advance of the contending waters, and whose rugged and broken outline forms a marked contrast to the more regular features of southern Britain.

The Physical Structure too of Scotland contrasts strongly with that of England. Built up of materials far older than the mass of the English rocks, the country presents an aspect of rudeness and wildness wholly unlike that of the bulk of the southern kingdom. In England the general level of the land is less than 500 feet above the sea, and the mountains that attain to 2,000 feet are few in number. In Scotland, on the contrary, no less than three-fourths of the entire surface of the land are covered by masses of gneiss, quartz, granite, and Silurian rocks, which as they tower into lofty peaks and ridges, and break into crags and precipices, form mountains that in some cases reach to nearly double the height of most of the chief English summits, or widen into moorlands that lie 1,000 feet or 1,500 feet above the sea.

Scotland consists in fact of two masses of mountains, each of which stretches across the country from sea to sea, and of a narrow plain that lies between them. The mountain masses strike across the land from west to east, having thus a direction almost at right angles to the main hill-range of England, the Pennine chain ; but they differ widely in extent from one another and from the valley which they enclose. The *High-lands*, which cover the northern part of Scotland, have an extreme breadth from sea to sea of 160 miles, and rise in their highest summits to 4,400 feet. The *Lowland Hills*, which occupy the south-ernmost part of the country, measure 125 miles ·across their broadest part, and rise in their loftiest mountain to 2,500 feet.

The *Lowland Plain*, on the other hand, which fills the isthmus by which the hill-masses of northern and southern Scotland are linked together, has only a breadth of from 30 to 60 miles, and a large part of it does not lie 100 feet above the sea. Though comparatively small in extent, however, the structure of its rocks has made of it the most important of the three divisions of Scotland. Belonging as it does to a somewhat later geological age than the mountains which shut it in on either side, it is mainly formed of the same carboniferous rocks as those which constitute the Pennine chain in England; and like the Pennine moors it is rich in coal measures with abundant stores of iron.

Industries.—The physical structure of Scotland thus presents great obstacles to agricultural industry. It is only possible to till the soil within the narrow limits set by barren mountain ranges, which in fact leave but a fourth of the whole land open to cultivation; while of this small proportion as much as one-half is most profitably given up to grass. In spite however of this scarcity of land for tillage, a careful and scientific mode of cultivation has made of the Scotch Lowlands some of the most productive districts in Britain. The mountainous tracts of Scotland too, though wholly devoid of mineral wealth, afford vast pastures, where a hardy race of sheep and cattle find grazing ground; and the trade in these cattle forms an important branch of Scotch industry.

To this may be added the great fisheries of the coast for cod, herring, haddock, and ling, and that of the mountain streams for salmon.

But the bulk of the wealth of Scotland comes from its manufactures and commerce, both of which are dependant on the abundant supply of coal and iron that lies in the centre of the country—in that part of it where they can be most easily and profitably utilised, where the climate is most temperate, and where river-estuaries running up into the heart of the coal-measures form water-ways which connect them with the sea. A variety of industries have

therefore sprung up on the coal-fields of this central
plain—the manufacture of linen, jute, cotton, wool,
silk, and carpets; soap-making, distilling of spirits
from grain, iron-works, ship-building, and the making
of machinery. A large and varied commerce necessarily follows this manufacturing industry—the
bringing in of cotton, silk, flax, jute, and other raw
materials, and the sending out of all these in their
manufactured state, with machines, worked-iron, and
coal.

Population.—In spite of these industries however the population of Scotland remains small, owing
to the wild and rugged character of the country. The
number of its inhabitants is but 3,360,018—that is,
little more than one-seventh the population of England, or only equal to 100,000 more people than
the inhabitants of London. Of this scanty population, too, but a very small portion is found in the
mountainous districts. These form for the most part
vast solitudes, with scarcely an inhabitant: they have
no towns, no villages, only a few scattered hamlets
lying in little river valleys, and a thin fringe of small
fishing-ports which border the sea-shore. The bulk
of the people is massed together in the central
plain, where they have been drawn by a more sheltered climate and the possibilities of trade and agriculture. As we have seen (p. 10) these dwellers in
the plain and southern hills of Scotland represent the
later English race which peopled Britain, while the
mountains of the north give shelter to the earlier Celtic
or Gaelic inhabitants of the land.

From this general survey of the country we may pass on to examine in detail the physical structure of Scotland, to study the disposition of its mountains and its lower grounds, the lie of its greater rivers, and the conformation of its coast. From thence we shall proceed to examine the political divisions into which its people have been grouped.

CHAPTER II.

THE whole geography of Scotland hangs upon the structure of its mountain masses, for covering as they do three-fourths of its surface, they form the leading physical features of the country. It is, in fact, by the strike of its chief mountain ranges, and by the grouping of their dependant hills and ridges, that the character of the land is determined, whether we consider its internal order, or the outer form of its coast line. We shall therefore first consider the disposition of the mountains of Scotland under the two great divisions into which they naturally group themselves—those of (I.) **the Highlands, and** (II.) **the Lowland Hills.** Lying outside these two groups there are indeed isolated ranges of hills of igneous or volcanic origin, but these are distinguished from the rest by extent, by position, and by the different character of their rocks, and thus lie apart from the mountain groups with which we have now to do.

THE HIGHLANDS.—The most extensive and important of these mountain masses is that formed by the *Highlands*, which occupy the whole of northern Scotland from the Atlantic Ocean to the borders of the - Lowland Plain. A line drawn across the island from Stonehaven to the Mull of Cantyre will indicate the southernmost boundary of this mountain region ; and along this line, from Stonehaven to the Firth of Clyde, the meeting-point of plain and mountain is marked by a barrier of heights so disposed as to form a continuous wall which separates the wild and mountainous region to the north-west of it from the low-lying and fertile lands to the south-east. This wall of heights is interrupted indeed by a number of openings formed where the mountain streams break from the hills into the plain; but these openings constitute only passes which lead into the interior of the Highland district, where behind the rocky barrier through which they pierce a vast tract of mountainous country extends across the island from east to west, and stretches northwards to the extreme limits of Scotland in broken and crumpled masses of gneiss, schist, quartz, granite, and other crystalline rocks.

General Features.—The mass of mountains thus formed covers an extent of country three times as large as Wales, and like the Welsh and Cumbrian Hills represents the remains of an ancient land which rose above the ocean long before the formation of the England we now see. The bulk of the Highlands, that is, the whole of its central district, is built up of clays, slates, and limestones of the Silurian age.

To westward this central mass is flanked by a belt of fundamental gneiss which is nowhere else to be found in the British Isles, but which is in fact the most ancient rock which they contain, and that on which all the rest are built up. To eastward, on the other hand, it is bordered by a strip of Old Red sandstone, which extends along the shores of the Moray Firth and forms a fringe of fertile land by the sea-coast : this belt of sandstone, the latest formation of the Highland district, is as old as a great part of the hills of Devon and Cornwall.

But though the Silurian rocks of the Highlands belong to the same age as those of Wales, they differ from these in having 'undergone changes in structure which have given to them a crystalline character, so that their schists, mica-slates, flagstone, quartzose, and gneissose rocks are altogether unlike the materials which make up the Welsh and Cumbrian mountains. The great variety of the rocks which thus constitute the Highlands produces a corresponding variety in their scenery. The harder stones are built up into mountains with rude and craggy outlines, and the landscape thus formed is of a wild and savage character, as where quartz rocks tower up into lofty conical peaks of white and grey stone, or where masses of gneiss break into serrated ridges with little cones and spires and abrupt angular faces of cliff. The softer rocks, such as clay-slate, are worn by weather into rounded hills with smooth slopes covered with bog and heather, and present a tame, monotonous landscape in spite of the great height to which they often attain.

On the other hand, granite will assume many different forms : sometimes it stretches for miles in lofty undulating moorlands which never rise into a hill, while at other times it breaks into craggy precipices and perpendicular walls of rock, or is massed into great dome-shaped mountains. Everywhere, in short, the scenery is determined by an endless variety of rock-structure, with its varying susceptibility to the influences of weather and to the changes brought about by the constant action of water.

Glenmore.—The long succession of moorland and mountain thus formed is broken by a remarkable fissure which strikes midway across the Highland district from sea to sea in a line parallel to its south-eastern boundary, and which is called *Glenmore*, or the Great Glen. The length of this gorge from the *Inverness Firth* to *Loch Linnhe* is 100 miles, while between these openings to the sea its broadest part scarcely exceeds a mile. Steep mountains shut in the fissure on either side, and in its depths lie a series of lakes separated by tongues of land raised less than 50 feet above the sea-level, through which the *Caledonian Canal* has been cut, joining lake to lake and thus opening a waterway from the North Sea to the Atlantic Ocean.

Grouping of the Highland Mountains.— One of the most important features in the geography of the Highlands is the *irregular disposition of their mountain masses*. The whole of the Highland district is made up of lofty reaches of moorlands broken by confused masses of hills which never unite in such regular order as to constitute a distinct central range, and never link themselves together into a definite mountain chain. It is true indeed that they form a multitude of short ridges and walls of rock, but in comparison with the great mass of the mountains these are of trifling extent and do not afford landmarks of real consequence.

It is important to grasp clearly this leading fact in the structure of the Highlands, so as to avoid falling into a confusion of geographical terms and names. In the attempt to reduce to order so vast and irregular a mass of heights it has been the usual practice to describe a part of it under the name of the *Grampian Mountains*—a name which is inaccurate and misleading for two reasons; first, because it implies continuous ranges of hills which do not in fact exist ; and secondly, because it has been loosely applied to so many parts of the mountain mass as to have no longer a clear geographical meaning ; for while it is considered by some to refer to a belt of high ground which extends across the centre of the Highland district fron Ben Nevis to Stonehaven, by others it is limited to the south-eastern heights which border the Lowland Plain. Under these circumstances it is safer to dismiss the name altogether from our minds. But though we can only realise the true character of the Highland hill-masses by recognising their confusion and irregularity, we may still discern in their disposition an underlying order if we follow carefully the natural lie of the ground.

The Waterparting of the Highlands—For in spite of the absence of mountain ranges, there is one very important landmark in the Highlands which must form the starting-point for any study of their geography.

This is a *belt of persistent high ground* which traverses the whole of the Highland district from Cape Wrath to Loch Lomond, and which at once indicates the general strike of the mountain masses, and forms the great wind and water-parting of the country, since it severs between the districts that lie to east and west of it, and parts the rivers which fall into the Atlantic from those which fall into the North Sea. From either side of this central high ground proceed a number of dependant ridges and groups of hills, which strike to east and west in parallel lines, and are parted by deep lateral glens or by wider straths or river valleys. This peculiar formation, with its rough resemblance to the skeleton of the back and ribs of an animal, gave to the central ridge in old days the name of *Drumalban*, or the *backbone of Alban* or Scotland. This name, however, is no longer used; and to us the long irregular succession of moor and mountain which strikes down the centre of the Highlands is better known as the *Waterparting* range of the country, since it is in these central heights that the sources of the chief Highland rivers are found, and it is the lie of these river sources that gives us the best means of tracing the general direction of the heights.

The belt of high ground thus indicated is of varied character. Sometimes isolated mountain masses lie directly in the course of the waterparting line; but these again give place to lofty moorlands that stretch for leagues in monotonous wastes of heather and peat broken only by knolls and crags of rock, and flanked by mountains whose rugged peaks and serrated ridges rise to east and west out of the desert mountain plain on which they stand. Still it is possible by observing the succession of the more important among these heights to trace clearly the general direction of the backbone of the Highlands.

Beginning in the extreme north with the summit of *Ben Hope*, over 3,000 feet in height, we pass southward to the yet greater masses of *Ben More* and *Ben Dearig*. These three chief mountains are linked together by a long succession of moorlands which form some of the wildest and most savage scenery in Scotland. With their bleak stretches of heather broken by dark masses of bare stone, by high hills with craggy outlines, by deep clefts and defiles, and by precipitous ridges of rock whose crests are notched like the edge of a saw, they constitute a vast mountain wilderness almost without inhabitants. As they stretch southward beyond Ben Dearig they pass into a monotonous waste of heather known as the *Dirie More*, beyond which rises a mountain yet loftier than all the rest, *Ben Attow*, whose height of 4,000 feet is the greatest in all the Northern Highlands. A little further to the south the fissure of Glenmore strikes across the mountain mass, but on the southern side of this depression the line of high ground is again taken up, and is carried between the districts of *Lochaber* and *Badenoch* in reaches of bare rock and bog till it widens into the *Moor of Rannoch*. This

moor forms an open level plain which stretches for
leagues at a height of 1,000 feet, unbroken by a hill,
without tree or shrub of any kind, and varied only by
immense tracts of bog. From its southern border
are thrown out long reaches of undulating tableland
interrupted by rugged hills, which stretch as far as
Loch Lomond, and terminate on the border of the
Lowland Plain.

These central high grounds extending from Glenmore to Loch Lomond, are flanked to right and left
by steep hill ranges and isolated mountain masses.
To the westward towers the loftiest mountain of the
British Islands, *Ben Nevis*, a solitary upthrow of trap
or volcanic rock, 4,400 feet in height. A little to the
south of it rises *Ben Cruachan*, a mass of granite 20
miles in circumference and 3,670 feet high. Just
opposite to Ben Cruachan to the eastward lies a magnificent group of lofty summits and ridges, the mountains of *Breadalbane*, which extend from the Moor of
Rannoch on the north to the Lowland Plain on the
south, and thus form the termination of the entire
series of heights that constitute the backbone of the
Highlands. This Breadalbane group consists mainly
of short, sharply-defined ridges thrown out to the
eastward, and marked by important mountains, such
as *Schiehallion, Ben Lawers*, and *Ben More*. Like
the neighbouring hill-masses, *Ben Voirlich, Ben
Ledi*, and *Ben Lomond*, these ridges are severed
from one another by deep lateral valleys enclosing
lakes or mountain streams, and end abruptly as they
reach the Plain so as to present the appearance of a

barrier stretching obliquely across it. They form, in fact, a part of that wall which we have described as separating the Lowlands from the Highlands (see p. 252), a wall which we now see to be formed, not by a continuous mountain range, but by the sudden termination of a number of branches, which are one after another cut short as they abut on the low country, while from between them break the mountain streams whose valleys form the only passes into the Highland country.

We have thus roughly traced the course of that belt of high ground which forms the backbone of the Highlands, with the chief mountain masses that are immediately connected with it. This central ridge, running as it does from Cape Wrath to Loch Lomond, and throughout the whole of this distance lying near the western sea, breaks up the Highland district into two wholly distinct parts, one of which extends along the western, the other along the eastern coast. Both of these districts are traversed by dependent ridges and outlying hills which proceed from the high ground of the water-parting. Deep lateral glens and straths part these dependent ridges and enable us to determine the general lie of the hill-masses, and by following them we find that the main direction of these irregular groups of heights is from east to west, or at right angles to the central high grounds. But with this general resemblance in structure, the eastern and western districts of the Highlands differ from one another not only in size, but also in character.

The Western Watershed.—That part of the Highlands which forms the watershed of the Atlantic is a mere strip of land composed in great part of those rude rocks of ancient gneiss which lie along the north-western coast, building up a succession of bare rough hills and low headlands, utterly barren and treeless, without even a patch of green vegetation to relieve their cold grey heights and gloomy reaches of bog.   Behind these rise irregular but majestic pyramids of sandstone in huge isolated masses 2,000 or 3,000 feet high, with sides as steep as a wall, scored with deep rifts and fissures, far grander in colour and form than the older gneiss rocks, but quite as naked and barren.   There is no part of the massive mountain line from Cape Wrath to the Firth of Clyde, whether it consists of these craggy rocks, or of tame rounded hills of softer clay-slates, which is not penetrated by inlets of the sea, scored by glens and gullies, and cleft by the channels of mountain torrents, or by lakes long and narrow like rivers.   The torrents make their way into the sea by deep cuttings between walls of rock; and sometimes it happens that the slight barrier which severs a lake from the ocean is overflowed by the waves, and then a *sea-loch* is formed, like a narrow arm of the sea penetrating into the land.   By these sea-lochs the western coast of the Highlands is broken into a long succession of irregular peninsulas, such as *Glenelg, Morven, Appin, Lorn, Knapdale* and *Cantyre,* and *Cowal.*   To the north of Glenmore, the general lie of the glens and their openings to the sea, as well as of the

peninsulas and islands of the coast, trends in a north-
westerly direction; while to the south of Glenmore
on the other hand the whole direction of the coast
and of its islands is to the south-west.

The Eastern Watershed.—The district which
lies to eastward of the waterparting, and forms the
watershed of the North Sea, is of far greater extent
than that which lies to the west, and differs as greatly
from it in character. Instead of short rocky ridges
it is occupied by extensive moorlands, by irregular
heights falling slowly to the shore, and by lines of
hills occasionally linked together so as to approach
the character of mountain ranges. The streams, instead of running in short parallel glens, follow long
winding courses in broad straths or valleys between
the masses of out-lying hills, and gradually gather
together so as to form by the union of their waters
important river-basins. As they approach the sea
they enter on the belt of low sandstone that lines
the shore, and being watered by their streams makes
a rich agricultural district; and in this softer soil they
cut broad estuaries by which they empty themselves
into the ocean and thus form important harbours.
Rude and wild therefore as the whole of the Highland region is, the eastern watershed is the more
fitted for habitation by its extent, the character of its
soil, its river-basins, and its sea-ports.

A great inlet of the sea, called the *Moray Firth*,
has cut its way deep into the centre of this eastern
district, in the midst of the low land where most of
its river estuaries meet, and in which lies the north-

eastern opening of the valley of Glenmore. By this great bay, and by Glenmore itself, the eastern Highlands are broken into two large peninsulas which are thrown out in a north-easterly direction from the water-parting range. In both these peninsulas alike the greatest heights lie to westward near the backbone of the country, and the ground gradually falls to the eastward till it sinks at last into the plains that border the coast.

(1.) The district which lies to the north of the Moray Firth consists mainly of the peninsula of *Caithness*, a tract of moorland country wholly formed of Old Red Sandstone. Its general character is that of an open treeless plain of no great height, where reaches of bog and moss alternate with stretches of pasture land well stocked with cattle, or with level cultivated plains near the sea-shore. To the westward, however, the sandstone and other rocks of Caithness give place to higher and more rugged tablelands of slate and granite, such as those which form the mass of the northern Highlands. Reaching to an elevation of 1,000 or 1,500 feet, these constitute a barren mountain wilderness, extending westward as far as Ben Hope and Ben More, and broken by isolated summits such as that of *Ben Klibreck*, over 3,000 feet in height.

The peninsula of Caithness is bounded on the south by the *Morven Hills*, and by the *Ord of Caithness*, which falls in a sharp descent of 700 feet to the sea-shore ; from this point along the western border of the Moray Firth the country is traversed by irregular mountain ridges which extend almost to the coast, leaving but a narrow fringe of land, from one to ten miles wide, fit for tillage. A larger tract of arable ground is formed by a low spit of soil that projects into the sea between the Dornoch and Moray Firths, and forms the fertile *plain of Cromarty*, a plain which extends some little way inland along the shore of the Cromarty Firth, but is bounded to the westward by the hilly country of *Ardross*, and by the massive mountain of *Ben Wyvis*, which rises at the inner end of the Firth to a height of 3,700 feet. From Ben Wyvis southwards to *Mealfourvounie*, on the border of Glenmore, the country is occupied by outlying hills of the water-parting range, which for the most part form parallel ridges enclosing glens like those of the western coast.

(2.) The district which lies to the south of the Moray Firth is more important in size and character. Its geography is marked by larger features, its mountains are higher and more massive, its valleys larger and more fertile, its river-basins more extensive. The heights in this district too present somewhat less of the appearance of confused masses of hills; for here a system of alternate river valleys and lines of high ground all trending in the same direction have more a look of mountain ranges than is found elsewhere in the Highlands.

It is, however, by the valleys which limit the hill-ranges that their outline is really defined, and it is therefore necessary in the first place that these valleys should be distinctly marked. They are four in number. *Glenmore*, with its long narrow line of lakes,

forms the north-western boundary of the district;
Strathmore, which is really a part of the Lowland
plain, limits it to the south-east; while between
these outer boundaries lie long straths or valleys
marked by mountain streams, such as the *valley of
the Spey*, and the *valley of the Dee*, which form well-
marked boundaries to the irregular hill-masses that lie
between them.   In this manner the whole district is
broken up into three main belts of high ground.

(*a.*) Glenmore and Strathspey lie in parallel lines,
both having a north-easterly direction. Between them,
and running in the same direction, is the range of the
Monadh-Leadh Mountains, a chain of hills, heavy,
rounded, and barren, but without grandeur of form.

(*b.*) The mountains enclosed between Strathspey
and the valley of the Dee are loftier and more irregular.   They begin in the south-west, where these
two valleys almost meet, in the massive group of the
Cairngorm Mountains, scarcely lower than Ben Nevis,
and higher than any other mountain in Britain. Three
summits of this group, *Cairntoul, Cairngorm*, and *Ben
Macdhui*, rise over 4,000 feet in height, and *Ben
Avon*, a little to the eastward, is but slightly lower.
From this point, however, the mountains rapidly fall,
sending out long lines of irregular hills which branch
out so as to cover the broad tract of country that
is included between the valleys of the Spey and
the Dee on either hand, and terminate in the low
tumbled ground which forms the peninsula of *Buchan*,
one of the richest grazing countries in Scotland.

(*c.*) To the south of the valley of the Dee lies a

much more marked and important belt of heights, composed of a continuous chain of lofty summits that run east and west from Loch Ericht almost to the North Sea, and which nowhere sink below 2,000 feet in height. The western summits of this chain, such as *Ben Dearg,* look northward over the upper valley of the Dee to the Cairngorm heights, and thrust out to the south past *Ben-y-Gloe* long spurs called the *Athol Mountains* that run down to a gorge which severs them from the Breadalbane hills. To the east of Ben Dearg the range is prolonged by the heights of *Glas-Mhicl, Loch-na-gar,* and *Mount Bat-tock.* Here the long southern slopes of the hills are known as the *Braes of Angus,* and as they sink to the plain of Strathmore where they touch the Low-land valley, they form the north-eastern part of the great Highland wall which we have before described as overlooking the Central Plain (see p. 252). The continuous wall of heights formed by this mountain chain cuts off all communication between the districts that lie to north and south of it, save by passes formed at its two extremities which constitute the only routes that connect northern and central Scotland. One of these routes leads by the sea-coast from Buchan to Strathmore: the second lies to westward, opening out of Strathspey into the valley of the Tay, and is formed by the *Pass of Dalwhinnie,* the deep glen or gorge of the river *Garry,* and the *Pass of Killiecrankie.* This long valley marks in fact the division between the chain of heights that we have been following and the group of the Breadalbane Mountains, as it is closely

shut in to east and west by the outliers of these
two hill-masses.

Highland Lakes.—We have now gained a general
view of the disposition of the Highland Mountains ;
but one great feature of this district yet remains to be
spoken of. This is the system of lakes that penetrate
the glens by which the mass of the mountains is inter-
sected, or that are gathered in hollows sunk in their
broad moorlands. The lakes of Scotland taken together would form a small sea of 500 square miles in
extent, and almost all that are of any size lie within
the Highland district. For the most part they are
gathered into its *western* half, where narrow glens are
shut in between steep mountain ridges, and where the
lakes take almost the form of rivers, being generally
many miles in length and scarcely one in breadth. In
the Highlands to the north-west of Glenmore there
is scarcely a gorge without its lake, large or small ; and
these lie ranged in nearly parallel lines along the
eastern and western slopes of the ridge of high ground
which runs from Ben Hope to Morven. *Loch Shin*,
the second greatest lake of the north, is 17 miles long,
and but one mile wide : it lies between low moory
banks, and forms one of a long chain of waters which
rival those of Glenmore, and almost cut a channel
across northern Scotland between Ben Klibreck and
Ben More. *Lake Maree* drains its waters into the
Atlantic to the south-west of Ben Dearig ; it is greater
in extent than Loch Shin, and of greater beauty, with
its group of wooded islands and its surrounding walls
of wild and massive mountains. To the south of Ben

Attow the lakes become yet more numerous, though smaller. Those on the eastern slopes of the central ridge of high ground drain their waters into Glen-more, as *Loch Quoich, Loch Garry,* and *Loch Arkaig,* while those on the west are connected with the Atlantic and are almost sea-lochs, such as *Loch Morar* and *Loch Shiel.* Glenmore itself is, as we have seen, occupied by a continuous chain of lakes. The greatest of these, *Loch Ness,* is twenty-two miles in length : the lesser ones, *Loch Oich* and *Loch Lochie,* are but five and nine miles long. *Loch Eil* and the *Firth of Inverness,* which form the extremities of the glen, open out into the ocean and are therefore sea-lochs.

But the district which is especially marked by the number, the extent, and the magnificent scenery of its lakes, lies in the southern part of the Highlands, in that mass of mountains which includes Ben Nevis, Ben Cruachan, and the Breadalbane Mountains, and which forms the extremity of the backbone of the Highlands. Within this group of mountains the lakes are ranged round a central table-land, whence they radiate outwards like the spokes of a wheel, filling the deep gorges which sever the lateral ridges that are thrown off from the central heights, or lying in the moorlands that stretch between isolated mountain summits. Their arrangement thus resembles that of the lakes of the Cumbrian mountains, though on a much vaster and more splendid scale. Beginning in the north with the smallest of them, *Loch Laggan,* we pass by *Loch Ericht, Loch Rannoch, Loch Tay, Loch*

Earn, Loch Katrine and its chain of small dependent lakes, to *Loch Lomond.* To the westward, *Loch Awe* stretches from the foot of Ben Cruachan through a distance of twenty-three miles, forming the last of the group of inland waters ; while to north and south of it a number of sea-lochs, *Loch Long, Loch Fyne, Loch Etive,* and *Loch Leven,* complete the circle.

Of all the Scotch lakes, Loch Lomond is the largest and the finest in scenery : it is twenty-four miles in length, and has the form of a thin wedge driven up into the heart of the mountain masses, while its broader southern extremity, dotted with wooded islands, opens to the Lowland plain. *Loch Katrine* is smaller, being only eight miles long, but is famous for the romantic beauty of its scenery, and the grandeur of the approach to it through the deep gorge of the *Trossachs.*

CHAPTER III.

IF we now pass from the Highlands to a similar survey of the Lowland hills, we shall trace the points of resemblance and of difference which exist between the mountains of northern and southern Scotland.

THE LOWLAND HILLS occupy the southernmost part of Scotland, filling the whole of that broad tract of land which is narrowed to the north by two arms of the sea, the Firths of Forth and Clyde, and to the south by the Solway Firth. From the cliffs of *Portpatrick* in the south-west to the rocky shore by *St. Abb's Head* in the north-east they stretch in a continuous belt of high ground that traverses the whole breadth of Scotland from sea to sea, and which formed of old its great defence against England to the south.

The north-western face of this range overlooks the Lowland plain, from which it rises in a wall of heights as sharply marked as that of the Highlands which

front it from the other side of the great valley. From
Girvan on the North Channel to the upper valley of
the Nith this north-western boundary of the Lowland
hills is shown by a line of comparatively well-
marked heights. The border line is less clearly defined
between the valleys of the Nith and the Clyde, owing
to the broken and hilly character of the ground in
front; but it is once more taken up by the chain
of the *Lammermuir Hills* as they stand out from
the plain with steep bare front like a wall of sea-
cliffs. The south-eastern boundary of the Lowland
hills is less distinctly marked than that of the north-
west. Long lines of hill and moorland alternating
with river valleys thrown out to the southward link
their central heights with the *Cheviot Hills* on the
English border, or sink in quiet pasture-covered
slopes to the shore of the Solway Firth.

**Contrast between the Lowland Hills
and the Highlands.**—(1.) Within these narrow
limits lies a tract of mountains of *small extent*,
if we compare them with the mass of the Highlands.

(2.) Besides their difference in size there is a difference too in the *character of their rocks*. While
both mountain-groups belong to the same far-distant
Silurian age, the Lowland hills are more simple
in structure; the masses of shale, grey-wacke, and
limestone, which have been changed or metamor-
phosed in the Highlands, have remained unaltered
in the Lowland Hills, where they preserve the
character of the Welsh and Cumbrian rocks, and

are much less broken by blocks of granite, trap, and other igneous rocks.

(3.) The comparatively simple formation of the Lowland hills has produced in them a *scenery of a much quieter and more monotonous character* than that of the northern mountains. Their greatest heights fall short by nearly 2,000 feet of the chief summits of the Highlands, and never present the same rugged and savage appearance. With their broad flattened hill-tops, their grassy slopes, smooth, green, and shrubless, and their deep narrow valleys, they have no likeness to the serrated ridges, the crags and precipices, the masses of forests, and the broad straths of the Highlands. Consisting as they mainly do of a lofty and extensive table-land, the Lowland hills form in fact an undulating pastoral country traversed by a multitude of streams which hollow out deep river valleys, and so cut the table-land into a series of ridges or into irregular hill masses. In some parts indeed the heights are wild and rugged, deeply trenched with glens and gullies, and scarred with precipices and faces of bare grey crag. But for the most part the summits of the hills join on to one another and form a level plain, it may be 1,500 or even 2,700 feet in height, without a shrub or tree, but uniformly covered with coarse grass or heather. From this upper ground the hill-sides fall in quiet monotonous curves, and present a continuous succession of pasture-covered slopes dipping to the narrow valleys that form their river channels.

(4.) The Lowland Hills differ also from the Highlands by the fact that they more nearly approach the character of a *continuous range of heights.* This range strikes across the island from Portpatrick to St. Abb's Head, and its central ridge is marked by the sources of the rivers which are thrown down its northern and southern slopes.

Grouping of the Lowland Hills.—While however the whole mass of the Lowland hills thus shares in the same general features of structure and scenery, yet by the natural lie of the ground it is broken into two nearly equal parts, each marked by special peculiarities. The division between them is formed by the valley of the *Nith*, as it strikes from north to south across the centre of the hill-country, and parts between the districts which lie to west and east of it, districts which are distinguished from one another somewhat in the same way as the Highland districts which lie to west and east of the valley of the Garry. The *south-western* half, from Portpatrick to Nithsdale, is very much broken by irregular masses of mountains, among which are to be found some of the highest summits of the Lowland hills : numerous lakes and tarns are scattered among the heights, but there is scarcely a river of any size. The *north-eastern* half, on the contrary, from Nithsdale to St. Abb's Head, consists of long lines of hills bound together in continuous ranges; it contains hardly any lakes, but is traversed by a complete system of streams which unite their waters to form extensive river basins thus, and fall into the sea as large rivers. We must briefly survey both these districts.

(1.) The *uplands of the south-west* take their rise
on the shores of the North Channel in long reaches
of rough moorland varied by a succession of little
rounded hills and by mountain tarns. This moorland
rises on its eastern side into a group of mountains
whose chief summit, *Merrick*, 2,764 feet above the
sea-level, forms the highest ground in southern
Scotland, while in the grandeur and desolation of
its scenery, in its crags and precipices and deeply-
scored gullies, it almost approaches the mountains of the north. The *Rhinns of Kells*, only a few
hundred feet lower, carry on this series of heights
in a north-easterly direction to the point where
Black Larg rises above the valley of the Nith.
These masses of hills form the crest, as it were, of
the uplands along their north-eastern side, and
from them a tract of country slopes southwards,
which breaks into peninsulas thrown out into the
Solway Firth. This tract consists of long stretches
of lower hill and moorland cut into ridges by little
valleys, and everywhere presenting the same green
treeless slopes. To the south and south-west of
Merrick stretch the *Galloway Hills* in long succession on either side of Wigton Bay. To eastward of
these the country is more open, with broader river
valleys, broken however at the mouth of the Nith
by a solitary upthrow of granite, *Criffel*, which rises
to a height of 1,867 feet.

(2.) The uplands of the north-east, from Nithsdale
to St. Abb's Head, begin in the *Lowther Hills*, which
front Black Larg on the opposite side of Nithsdale,

T

and which resemble the south-western mountains in
their rocky precipices shutting in narrow defiles,
and in the wildness of their scenery.   To eastward of them, beyond the vale of Annandale, lies a
second group of mountains which forms the most
important centre of this district, answering to that
of Merrick in the south-west.   Scarcely lower than
Merrick,   these   mountains   of *Broad Law*, *Dollar
Law*, *White Coomb* and *Hart Fell*, are ranged in a
series of heights which determine the river-system of
the surrounding country, heights from which all its
great valleys diverge to north and south and east, and
which form the connecting link between two separate
belts of uplands that branch out from this point, one
north-east to St. Abb's Head, the other south-east to
the Cheviots.   In these mountains the character of a
high table-land is strikingly shown.   Their smooth flat
summits are so joined together as to become an upland plain that mounts from 2,600 feet to a height of
2,754 feet in its loftiest reach of *Broad Law*, whose
summit, covered with heather and peaty moss, forms
a solitary moorland that stretches far and wide and is
as level as a racecourse.

The high table-land that thus runs north and south
from Dollar Law to Hart Fell is prolonged at either
end into two distinct ranges of hills severed from one
another by the broad valley of the Tweed.

(*a.*) To the north-east its line is taken up by the
Muirfoot Hills, and by the long chain of the
Lammermuir Hills, which stretch away to the shores
of the North Sea.   These last rise steeply from the

Lowland plain in cliffs of 1,500 or 1,600 feet in height, cliffs which form the edge of a high table-land like that of Broad Law, where the hill tops are linked together into a continuous moorland, smooth, green and treeless, whose southern slopes descend in long reaches of pasture land to the valley of the Tweed.

(*b*.) The line of hill and moor which bounds the opposite side of this valley extends to the south-east from the neighbourhood of Hart Fell, from whence by the heights of *Ettrick Pen*, *Wisp Hill*, and *Peel Fell*, it leads to the chain of the *Cheviot Hills*. This range of heights, itself an upthrow of vol-canic rock, is wholly distinct in character from the mass of the Lowland Hills. It stretches along the English border eastward to the North Sea, reaching in its chief summit, *Cheviot Peak*, to a height of 2,700 feet.

The Lakes of southern Scotland are of little im-portance as compared with those of the Highlands. The great bulk of them lie, as in the Highlands, to the westward, and have the same long and narrow form ; but of the numerous sheets of water which vary the south-western moorlands, there are but two that are more than mountain tarns, *Loch Doon* and *Loch Ken*, and these have a length of only five or six miles, and a breadth which in Loch Doon alone reaches a mile. In the north-eastern half of the Lowland Hills there is but one small lake, *St. Mary's Loch*, which lies under the heights of Dollar Law, and pours its waters through the river Yarrow into the Tweed.

We have now gained a general idea of the structure and arrangement of the two mountain masses of Scotland, and have seen the character of wildness and desolation which they give to the country of which they form so large a part. But between these mountain groups, and shut in by steep walls of rock, lies a tract of land distinguished by very different features, the Lowland plain; and it is this plain, with the outlying hill ranges that intersect it, to which we now pass on.

Highland Mountains.

	Feet.		Feet.
Ben Nevis	4,406	Ben Dearig. . . .	3,551
Ben Macdhui . . .	4,295	Ben Dearg	3,550
Cairntoul	4,285	Glas-Mhiel . . .	3,500
Cairngorm	4,090	Ben More	3,281
Ben Attow	4,000	Ben Lomond . . .	3,192
Ben Wyvis	3,720	Ben Voirlich . . .	3,180
Ben Lawers . . .	3,984	Ben Klibreck . . .	3,157
Ben More	3,818	Ben Hope	3,060
Ben Cruachan . . .	3,670		

Lowland Hills.

Merrick	2,764	Hart Fell	2,651
Broad Law. . . .	2,754	Culter Fell	2,454
White Coomb. . .	2.695	Lowther Hills. . .	2,377
Dollar Law . . .	2,680	Ettrick Pen . . .	2,269
Rhinns of Kells . .	2,668	Black Larg. . . .	2,231

The Highland Lakes.

Lakes.	Length in Miles.	Breadth.
Loch Lomond	24	7
Loch Awe	23	1¼
Loch Ness	22	¼
Loch Shin	17	1
Loch Tay.	14	1½
Loch Ericht	14	—
Loch Maree	12½	3
Loch Morrer	11	1
Loch Lydoch	10	½
Loch Lochie	9	1¼
Loch Rannoch	8	1¾
Loch Earn	8	¾

The Lowland Lakes.

Loch Leven	4	3
Loch Doon	6	1
Loch Ken	5	½
St. Mary's Loch . . .	3	¾

CHAPTER IV.

BETWEEN the mountain masses of northern and southern Scotland lies a belt of low ground that stretches across the country from sea to sea, and which forms by the comparative mildness of its climate, by the fertility of its soil, by its mineral wealth, and by the advantages which it possesses both for internal traffic and for foreign commerce, the natural centre of its political and industrial life. The differences of structure that mark the many varied plains of England and give them those special advantages that distinguish them from one another are all united in the Lowland plain of Scotland; a tract which has at the same time coal-measures and manufactures like the basin of the Trent, level agricultural lands like the basin of the Ouse, facilities for commerce like the basin of the Mersey, and political importance like the basin of the Thames. The natural grouping of population in a district so favoured by nature has made this valley the most important part of the northern kingdom so far as the industrial and political history of its people is concerned.

In one important point however the Lowland plain of Scotland differs from most of the plains of England; *it does not coincide with any river basin.* It consists of a great trough sunk low between the mountain masses which border it on either hand, being shut in to the north-west by the Highland mountain-wall and to the south-east by the rise of the Lowland hills.   On two opposite sides it lies open to the sea, which penetrates far into the land in the wide bays known as the Firth of Forth and the Firth of Clyde, leaving but a narrow belt of land between their inner extremities.

The undulating country which lies within these limits, though for the most part not 200 feet above the sea-level, does not in the strictest sense of the word form a plain, for its quiet curves are in very many places broken by masses of igneous rock rising abruptly from the low ground, and even by long chains of hills of old volcanic materials.   As these hills, interrupting the natural lie of the ground, break the plain into separate divisions, and as they form striking features in its scenery, we must first trace their general arrangement.

Its Hills.—The main belt of these igneous rocks and hills stretches across the entire breadth of the plain from the Firth of Tay to the Firth of Clyde in a direction exactly parallel to the well-marked boundary line of the Highland mountains. Its north-eastern extremity consists of two lines of hills which border the shores of the Firth of Tay.   To the north of this bay lie the *Sidlaw Hills,* a chain of heights rising in parts to

1,400 feet, while to the south are ranged the *Hills of Fife* and the *Lomond Hills*. These are prolonged to the south-west in the chain of the *Ochill Hills*, whose chief summit, *Ben Cleugh*, is 2,359 feet in height and forms the loftiest mountain of central Scotland. Yet further to the south-west lie the range of the *Campsie Fells*, 1,500 feet high; while beyond these the *Kilpatrick Hills*, the hills of *Renfrew*, and those of *North Ayrshire*, border the shores of the Firth of Clyde. Even across the Firth the same line of volcanic rocks can be traced in the islands of Cumbrae, Bute, and Arran, and the extremity of the peninsula of Cantyre.

A second range of these volcanic rocks makes a belt of high ground that runs from the hills of Renfrew to the south-east beside the valley of the river Clyde, and parts this stream from the coalfields of Ayrshire on the west.

A third line of hills marks the south-eastern limits of the plain. These heights are very irregular, sometimes formed of sandstone, sometimes of igneous rock, and lie in scattered groups in front of the Lowland Hills. The chief among these are *Cairn Table*, *Tinto Hill*, a solitary mass of igneous rock 1,335 feet high, the *Pentland Hills*, which reach to 1,800 feet, and the crags and sharply-marked heights that break the coast of the Firth of Forth, *Edinburgh Rock;* near it *Arthur's Seat;* and at the opening of the bay, *North Berwick Law*.

We see therefore that the Lowland valley is far from being strictly a plain—that is, a flat expanse of land— for it is always undulating, and generally much rougher than the low grounds of England. There is, in fact, only one true bit of plain in all Scotland, that which extends along the Forth below Stirling. Indeed, since the igneous hills which cut up the Lowland valley are never out of sight in middle Scotland, they go far to give it a mountainous character. The long lines of heights formed by the Kilpatrick Hills, the high table-land of the Campsie Fells, and the Ochill Hills, are, it is true, less bold in outline than the ridges of the Highlands, yet they cut off the central levels from the Highland country far more effectually than the moun-tain-wall of the Highlands themselves, and form one of the most striking features of the Lowland Plain. There is no point in which this valley is so unlike the plains of England as in this constant presence of hills.

In these heights of the Lowland plain, occupying as they do three sides of an irregular square, we have a rough framework which breaks the plain itself into clearly-marked divisions of fertile and well-watered land. The separate districts so formed, however, are not in any case made by a river-valley: it is true that they are traversed by a multitude of rivers large and small, but these streams flow across them in various directions, not running parallel to the line of the hill-ranges, but breaking through the heights at right angles as they pass from one belt of low country to another.

(1.) The first tract of low land consists of a belt of Old Red sandstone which lies in a long valley shut in between the line of the Highland mountains and the range of hills that extends from the Ochills to the

Sidlaws. The larger part of this valley, from the shores of the North Sea to the west end of the Ochills, is known as *Strathmore* or the *Great Valley*, for nowhere in Scotland is there so extensive a reach of perfectly level fertile soil as in this plain, 90 miles in length, and broadening from its north-eastern end, only one mile wide, to a width of sixteen miles in the south-west. It is traversed by a number of rivers, such as the Esk, the Isla, the Tay, the Earn, and the Forth, which all alike break from the Highland mountains, cross the plain at right angles, and cut their way through the hill ranges which bound it on the south-east.

(2.) A second tract of low land occupies the western sea-coast by the shore of the Firth of Clyde, and is known as *Ayrshire*. On the north-east it is parted by a low line of hills from the basin of the Clyde, and on the south-east it is bounded by the Lowland-hills stretching from Nithsdale to the sea. The crescent-shaped tract of country shut in within these limits is watered by a number of little streams thrown down from its eastern heights to the sea, the chief of which, the *Ayr*, divides it into two parts. To the southward lies *Carrick*, a district which really belongs to the Lowland hills, thinly-peopled and chiefly devoted to pasture-land. To the northward, in the district known as *Cunnynghame*, a belt of rich coal-measures extends to the estuary of the Clyde, and forms one of the centres of Scotch manufacturing industry. It is thickly crowded with towns and villages whose external trade is carried on through the ports of the Clyde.

(3.) The third and most important division of the

Lowland plain is that which is comprised between the Firths of Forth and Clyde, and which thus occupies the narrow peninsula that links together northern and southern Scotland.   To the north-west it is shut in by the Fife and Lomond Hills, the Ochill Hills, and the Campsie Fells; to the west by the low range that parts it from Ayrshire; and to the south-east by the belt of irregular country marked by Cairn Table, Tinto Hill, and the Pentlands.   The low land within these limits is traversed by two important rivers, which flow in contrary directions, and empty themselves into opposite seas; for the *Forth* passes into it through a gap in the hills to the north-west, while the *Clyde* enters it from the heights of the south-east.   Both rivers, however, alike form deep estuaries, which penetrate so far . into the heart of the land as to leave a distance of but thirty miles between them, and are connected by a canal which opens a passage for traffic right across the isthmus from sea to sea.   The lower basins of the Forth and the Clyde form some of the best land for tillage in Scotland, and agricultural industry is here carried to great perfection; but the main importance of this part of the Lowland plain lies in the fact that it is composed through nearly its whole extent of the same carboniferous rocks as those which constitute in England the Pennine range, and is thus enriched by beds of coal and iron ore that are only second in importance to those of Yorkshire.   These coal-measures, which can be worked throughout an extent of over 1,500 square miles, have become, by the advantages of their geographical position in the centre

of Scotland, and by the fact of their opening to the
sea on either side in magnificent river-estuaries, the
centre of a vast group of manufactures, while the
Firth of the Clyde forms one of the greatest seats of
ship-building in Britain.

With such a variety of industrial resources this
central portion of the Lowland Plain has become
the most prosperous district in Scotland. Though in
extent it is equal to only one-twelfth part of the whole
country, yet with its sea-ports, coal-pits, blast-furnaces,
woollen and linen manufactories, and fields of corn and
wheat, where every foot of ground teems with a wealth
unknown to the poverty-stricken mountains, it is able
to support, roughly speaking, one-third of the Scotch
people. Its importance as the centre not only of the
industrial activity but also of the political life of the
land is marked by the two cities that guard the isthmus
on either hand. To the eastward, among the agricultural plains that border the estuary of the Forth, lies
Edinburgh, whose geographical position, commanding
at once the estuary which faces towards Europe and
the high road from London and York to the north,
early made it the chief city in Scotland and the seat
of its government. To westward, on the other hand,
lies the city of *Glasgow*, whose outlook is towards
the New World. Situated among the coal-measures
of the Clyde, it owes its importance to the industrial
impulse given by the development of manufactures
and the opening of trade with America, and by population and wealth it has become the third greatest city
in the British Islands.

CHAPTER V.

General Characteristics.—Scotland, like the southern part of Britain, is traversed by an extensive system of running waters through which the excess of rain that falls on its mountains is thrown down to the sea ; but from the structure of the country its river system offers many points of contrast to that of England both in its physical characteristics and in the political and commercial results which follow from them.

(1.) As the mountain masses which form the reservoirs whence the Scotch streams are drawn are of much greater elevation and extent than those of England, the river springs lie as a general rule far higher than those of the English streams. For example, the headwaters of the Thames are but 300 feet above the sea, while the highest river source in Scotland, that of the Dee, lies at a height of 4,060 feet, or 1,000 feet higher than the loftiest mountain-top in England.

(2.) One result of this height of their sources is that the rapid fall of many of the Scotch streams gives them the character of mountain torrents rather than that of rivers, and renders them wholly useless for navigation during the greater part of their length. The rivers of Southern Britain, on the contrary, are often navigable to within a short distance of their sources.

(3.) Again, these mountain torrents are in most cases thrown down the hill-sides by the most direct way to the sea, and shut in as they are between ridges of rock and short chains of hills, they flow in parallel channels without taking up big affluents or uniting to form river-basins. Most of the Scotch river-valleys thus consist of comparatively narrow troughs whose boundaries run straight to the sea ; and the *glens* and *straths* so formed are wholly unlike the wide plains traversed by the streams of England, where the open and undulating character of the ground often throws a whole group of running waters into one system, and gives rise to river-basins of great extent.

(4.) The rivers which travel through the upper glens and defiles of the Scotch mountains are closely con-nected with the system of lochs by which these are penetrated, sometimes having their origin in some lake or tarn among the heights, and at other times passing through one or more lakes, and so linking them together. As we have seen, this is never the case in England save with the streams of the Cumbrian hills.

(5.) While the general character of the Scotch rivers hinders internal navigation, and prevents them from being of the same political and commercial importance

as those of England, yet, on the other hand, they
have one great element of commercial value in the
Firths or broad estuaries by which they pour themselves into the sea, and which form so marked a
feature in the coast line. The magnificent harbours
thus formed are among the most important in Britain ;
while by their geographical position in the heart of
Scotland, opening eastward to Europe and westward
to the New World, they have proved of the first consequence in the development of the national life of
the country and the growth of its material prosperity.

From these general remarks as to the characteristics
of the rivers of Scotland we must now pass on to a
general view of its whole river-system, and of the
more important among its streams.

The Waterparting of Scotland, or that belt of
rising ground which parts between the rivers that
flow into the Atlantic Ocean and the North Sea,
traverses the country from Cape Wrath in the north
to the Cheviot Hills in the south-east, from whence it
is continued in the same southerly direction through
England. In its course it traverses the three great
divisions of the country. For two-thirds of its length,
from the moorlands of the north to the shores of Loch
Lomond, it lies among the Highland mountains ; it
then strikes across the heart of the Lowland plain,
passing over low hills and undulating ground whose
rise is at times so slight as to be scarcely perceptible ;
and it finally enters on the Lowland hills, and is
carried to the English border.

Like the English waterparting, it is mainly formed

by a belt of high ground that lies nearer to the western than the eastern sea, and it therefore breaks the country into two unequal parts, the larger watershed lying to the eastward, and the lesser to the westward. These watersheds are also marked by general characteristics which correspond to those of southern Britain, save that in Scotland their differences are brought out with greater sharpness of contrast. (1.) The *greatest number of large rivers flow to the eastward :* the streams of the west are generally mere mountain torrents, and there is but one river of note, the Clyde, along the whole of the western coast. (2.) *The rivers of the west have a very short course* to the sea ; in fact, there are but two of them which have a length of over fifty miles. The eastern streams, on the other hand, have often a considerable distance to travel. (3.) *The western streams do not unite so as to form river-basins ;* save the Clyde, there is no river whose basin attains an area of 500 square miles. Some of the eastern rivers, on the contrary, drain very great tracts of country, amounting in one case, that of the Tay, to 2,400 square miles.

The Grouping of Rivers.—And not only is the character of the rivers affected by the difference of the districts on either side of the waterparting, but different portions of the waterparting itself have a different relation to the general river-system. The bulk of the Scotch rivers have their sources in the mass of the Highland mountains, which by extent and by the amount of their rainfall necessarily constitute the great reservoir from which the running waters of the country are supplied. In the Lowland Plain, on the other hand, not a single stream of any note takes its rise; while the Lowland Hills, of greater extent and elevation, give rise to a system of rivers, less indeed in mere size than that of the Highlands, but of nearly equal importance.

There are therefore but two main river groups—*the rivers of the Highlands,* and *the rivers of the Lowland hills.* These, with the special features that distinguish them, must be considered in order.

[A.] THE HIGHLAND RIVERS.—We have already seen that the great bulk of the rivers which take their rise between Cape Wrath and Loch Lomond are poured into the eastern sea. On the western side of the waterparting, where the mountain slopes are carved into lateral ridges of rock, the streams form short mountain torrents, which flow straight to the ocean through deep and narrow glens and terminate in sea lochs. But while on this side there is not a single large river, the case is very different if we look eastward. For here in straths and valleys rivulets gather into streams and streams into large rivers, so that the waters of the eastern slopes are poured through great river basins into the North Sea, and form the most important river group in Scotland. It is with this eastern group therefore that we have alone to do.

The rivers of the eastern Highlands are divided into two distinct groups by that belt of high ground which is thrown out in a north-easterly direction to the sea, past the Cairngorm mountains, and along the centre of the Buchan peninsula (see p. 264). To the northward

U

of it lie the *rivers which open into the Moray Firth;* to the southward *those which pour directly into the North Sea.* These basins are about equal in extent, though the groups of rivers which drain them are by no means equal in importance.

I. **The Rivers of the Moray Firth.**—The streams which drain the northern district of Scotland, the mountain wildernesses of Sutherland and Ross, are too small to need separate notice. Their course to the sea is short, and lies across wild, uninhabited moorlands; so that they never grow to any considerable size save where they unite as they approach the sea, and form estuaries which represent the drainage of large tracts of country. Such estuaries we find in the *Dornoch Firth*, the *Cromarty Firth*, and the *Inverness Firth.* This last, which opens into the central part of the Moray Firth, receives two distinct systems of waters. The *Beauley Firth* carries to it the streams sent down from the mountain-ridge to the north of Ben Attow; while the river *Ness* pours into it the drainage of Loch Ness, or the north-eastern half of Glenmore.

There is indeed but one large river which falls into the Moray Firth. This is the *Spey*, which lies between the Monadh-Leadh Mountains and the mountain chain reaching from Cairngorm to the hills of Buchan, and whose broad valley of Strathspey forms one of the most striking natural divisions of this district The river flows from a little pool, Loch Spey, which is situated at a height of 1,200 feet on the waterparting range in the district of Lochaber. Its

upper moorland course lies through a series of dark
stagnant lakes, but it breaks into a more rapid and
impetuous stream as it enters on its valley below the
Monadh-Leadh Mountains, with the hilly spurs of
the Badenoch district, the Cairngorm Mountains, the
Braes of Abernethy and the great summit of Corryhabbie shutting it in to the south-east.   Its rapid fall
of 1,200 feet in a length of 96 miles makes the Spey
unnavigable ; and though the area it drains is large,
nearly 1,200 square miles, the river varies constantly
in the amount of its water, owing to the changeable
character of the mountain streams by which it is
supplied.   In spite therefore of its size the Spey is of
no commercial value ; but in its basin, as affording the
only ground fit for tillage in this rugged region, are
gathered a number of agricultural villages.

   We need only mention some unimportant streams
which lie on either side of the Spey, and whose
channels have a parallel direction with it.   These are
the *Nairn* and the *Findhorn* to westward, and the
Doveran to eastward.

   II. **The Rivers of the North Sea**, as we
may term those which lie to the south of the Buchan
peninsula, have a general direction to east and south-
east, differing in this from the rivers of the Moray
Firth, whose openings commonly lie to the north-east.
Six rivers belong to this group: the two northernmost
of these, *the Don* and *the Dee*, lie close together, and
have parallel courses eastward, while the four remaining rivers, the *North* and *South Esk*, the *Tay*, and the
Forth, run in a more southerly direction.

The Dee and the Don have their headwaters at no very great distance apart in the chain of the Cairngorm Mountains, and empty themselves into the sea within two or three miles of one another. Their valleys, separated by lines of broken hills, lie in a due easterly direction, and form the trough or basin which we have marked as parting the succession of heights from Cairngorm to Buchan, and the mountain range between Ben Dearg and Mount Battock. The *Don* is the more northerly of the two streams : its headwaters lie at a height of 1,640 feet, and though the river falls from this elevation to the sea in a short course of fifty miles, its stream is even and little broken by rapids. The basin of the Don is small, being but 530 square miles in extent.

The *Dee* is larger than the Don, having a length of eighty-seven miles, and a basin of 700 square miles in area. Its springs are at a greater height than those of any other stream in Britain, lying far up among the mountains 4,060 feet above the sea, and the upper valley of the river penetrates into the very heart of the wildest hill-masses of Scotland. Along its left bank tower the summits of Ben Macdhui and Ben Avon, while to the right it is closely shut in by a continuous wall of heights which from Ben Dearg eastward to Mount Battock never sink below 2,000 feet. As the Dee dashes down the centre of this wild valley, broken by its rocky bed into a constant series of rapid falls, it passes through some of the noblest scenery in Scotland, in the midst of which are situated, by the banks of the torrent, Braemar, the visiting place of tourists, and Balmoral, the Highland home of the Queen. The scenery of the Mar district, or the lower valley of the river is less striking.   Here the Dee becomes more regular and tame, and finally empties itself into the sea at *Aberdeen*, the third greatest city of Scotland, and one of the chief trading ports of the eastern coast.

The four remaining rivers of the Highlands drain the great tract of country which extends from the Braes of Angus to the Firth of Forth, and from the North Sea to the Breadalbane Mountains.   Two of these rivers, however, the *North* and *South Esk*, are but short streams thrown down from the Braes of Angus, and are of little note save for the harbour of Montrose, which is formed at the mouth of the South Esk.

The drainage of this district is therefore carried to the sea mainly by the two remaining streams, the *Tay* and the *Forth*.

The **Tay** is a very different river from any of those we have yet considered.   Though it rises in the Highlands, the greater part of its course lies in the lowland country.   Its length of 130 miles makes it the largest river in the whole of Scotland ; the extent of the area which it drains, 2,400 square miles, is far greater than that of any other river basin in the country ; and the volume of water which it pours into the sea exceeds that of any stream in Great Britain.

In its upper course the Tay forms a rapid mountain torrent which rises to the northward of Loch Lomond and flows in a deep glen of the Breadalbane

Mountains below the heights of Ben Lawers and Ben More ; it then enters the waters of Loch Tay, from which it finally emerges after a distance of fourteen miles, and passes from the mountain defiles into the valley of the *Tummel*. Taking up this river, it bends with it to the south-east and strikes across the plain of Strathmore, where it is changed from a mountain stream to a tame and regular river of the Lowland Plain, winding among rich fields of corn and quiet pasture lands till it reaches the channel between the Sidlaw and the Ochill Hills, where it turns eastward to enter on the Firth or estuary by which it passes to the sea. The gap by which the river winds round the extremity of the Sidlaw hills is marked by a considerable city, *Perth*, a city which has become the capital of this district through the importance of its position as commanding at once the pass into the rich plain to the north of it, and the commerce of the Tay, which here becomes navigable as it widens to its estuary. The river now flows through a district of considerable activity, both agricultural and manufacturing ; its northern bank is bordered by a tract of richly cultivated land, the Carse of Gowrie ; while along both its shores lies the seat of one of the chief Scotch manufactures, that of linen and jute, of which the port of Dundee is the centre.

The Tay, as we have seen, lies in the midst of a semicircular basin of 2,400 square miles in extent, shut in to northward by the mountains of the Athol district and to westward by the Breadalbane Mountains. From both these mountain groups affluents are thrown down to swell the size of the central stream. The *Isla*, which joins it from the north, carries to it the waters of the northern range about Ben-y-Gloe. By its other tributaries it drains the Breadalbane Mountains. The *Tummel* is itself joined by the *Garry*, and these two streams pour into the Tay on its left bank the drainage of a system of lakes lying near the Moor of Rannoch. On its right bank, just as it enters the Firth, it takes up the *Earn*, which passes through Loch Earn in one of the southernmost glens of the Breadalbane district, and thence opens into the valley of Strathearn. Most of these tributaries, with many lesser ones, come to the Tay from the west in a direction parallel to its own upper valley, and the whole group of streams thus forms a series of parallel rivers which break from the lateral ridges of the Highland mountains as these ridges terminate abruptly on the Plain. By their lower valleys the streams form the boundaries of the mountain ranges to the southeast. By their upper valleys they form passes leading up into the heart of the Highlands (see p. 252).

The history of the **Forth** is almost the same as that of the Tay, save that it is on a much smaller scale. From its source to the beginning of its estuary this river is but sixty miles long, or little more than half the length of the Tay, and its basin of 645 square miles is but one quarter the size of the Tay basin ; its Firth, however, fifty miles long, forms an estuary of great consequence. Like the Tay, the Forth rises in the mountains near Loch Lomond, and breaking quickly from the high ground strikes across a belt of red

sandstone towards the igneous hills which bound it
to the south-east. Here it takes up at the mouth of
Menteith its chief affluent the *Teith*, which carries
to it the waters of a system of lakes lying round
Loch Katrine. Its passage to its estuary lies between
the extremity of the Ochill Hills and the Campsie
Fells. The gap formèd between these heights constitutes the one great pass that leads from southern
to northern Scotland, and is marked by the fortress
of Stirling, which from its rocky height at once
commands the entrance to the plain outstretched at
its feet and guards the head of the Firth. To eastward of Stirling the Forth enters on the eastern limits
of the coal-measures in the central plain, and from
this point its banks are lined for a short distance with
mining and manufacturing towns. To north lie the
coal-fields of Alloa, to south the ironworks of Falkirk.
Further eastward this mining district ceases ; and the
towns along the northern bank of the estuary, like all
those of the peninsula of Fife, which parts the Firths of
Forth and of Tay, are busied in the linen manufacture ;
while along its southern bank extends the agricultural
district of the Lothians, where the best tillage in Scotland is carried on, and in the midst of whose fertile
plains lies the capital of Scotland, EDINBURGH. The
situation of Edinburgh, near the banks of the Firth,
with its outlook towards the European mainland, early
marked out the city as a political and commercial centre
of the country.

[B.] Though THE RIVERS OF THE LOWLAND HILLS
drain a small extent of country in comparison with

those of the Highlands, they yet form a group of streams of considerable size and importance. The more undulating character of the ground facilitates the formation of river basins larger than any in the Highlands save that of the Tay; while the position of the waterparting gives to the streams a greater variety of direction than they can have in the north, and thus furnishes southern Scotland with good river estuaries and harbours on its western as well as on its eastern side.

From Culter Fell to Carter Fell, where the Lowland Hills touch the Cheviots, the line of waterparting no longer lies near the western coast, but is drawn almost across the centre of the Lowland Hills over Hart Fell and Ettrick Pen, dividing them into equal parts, which lie to north-east and to south-west. The rivers of this district therefore group themselves very differently from those of the Highlands, for as many important rivers are thrown down to the western sea as to the eastern, and it is in this quarter that the only western streams of any note in the whole of Scotland take their rise.

I. **The Rivers of the Eastern Lowlands** are all gathered up into one main stream, that of the Tweed, and by it poured into the North Sea. Rising in Hart Fell at a height of 1,500 feet, the **Tweed** flows northward, till through a gap between Dollar Law and the Muirfoot Hills it escapes to the east into a broad valley shut in between the Lammermuir and Cheviot Hills, where it winds through a plain of the richest soil in Scotland, watered by streams traversing it from the hills to north and south, and broken only by craggy knolls of rocks, the sites of ancient keeps of the Border warriors. The course of the Tweed through this rich valley was marked in old days by some of the most famous Scotch abbeys, Melrose, Dryburgh, and Kelso, which lay in well-chosen positions on loops and bends of the river, or at points where it takes up tributary streams; now, however, the only important places to note are *Galashiels* and other centres of the woollen manufacture which is carried on in Tweeddale. After a course of ninety-six miles the Tweed finally falls into the sea at Berwick, where it forms the boundary between England and Scotland. With its tributaries it drains an area of 1,870 square miles. From the Lammermuir and Muirfoot Hills it takes up on its northern banks the *Whiteadder*, the *Lauder*, and the *Gala*. From the south it gathers in a number of small streams that come to it through little mountain dales hollowed in the smooth sides of the hills—dales famous in poetry and the wild stories of Border life, and still notable for their quiet scenery of green slopes and pasture-covered hills. The *Ettrick* is thrown down from Ettrick Pen through the wild country known as Ettrick Forest; before joining the Tweed it takes up the *Yarrow*, a lesser stream which drains the waters of S. Mary's Loch. A larger stream, the *Teviot*, lies more to the south-west, and flows in a line parallel to the Cheviot Mountains to its junction with the Tweed at Kelso.

II. **The Rivers of the Western Lowlands** are broken into two divisions by the chain of hills that extends from the Lowther Hills to Merrick, and throws down streams (*a*) northward to the Firth of Clyde, and (*b*) southward to the Solway Firth.

(*a*.) There is but one large river which enters the Firth of Clyde ; and this river, the **Clyde**, forms the only stream of any consequence in western Scotland. It rises a little to the westward of the Tweed at almost the same height, 1,400 feet; it has the same length of course, ninety-eight miles, and a basin of nearly the same area, 1,580 square miles.   For a time its course lies parallel to that of the Tweed, in one of the bare green valleys peculiar to the Lowland Hills, and there is a point where the ground between the two streams becomes so low that it is needful to guard the banks of the Clyde lest it should flow into the valley of the Tweed.   Finally, however, the Clyde is turned to the north-west, and leaving the Lowland Hills breaks into the central plain near the town of Lanark by a series of rapid falls.   In its lower valley the character of the river changes ; after a short course among well-tilled fields and orchards, it passes into the most crowded district of the coal-measures, where they form the centre of the iron manufacture of Scotland.

The importance of the river lies in this lower valley, and in the wide estuary by which it opens to the sea.   On the river banks lies GLASGOW, the first port in Scotland and the industrial capital of the country, surrounded for many miles by a district which forms in fact but one vast town of mining works and factories for iron, silk, wool, and cotton. Even beyond Glasgow as far as its opening into the sea, the Firth of Clyde presents a ceaseless aspect of activity, for the whole of the ship-building trade of Scotland is carried on along its banks, and the ports and docks which line its shores are busied in the foreign trade with America, to which the towns of the Clyde owe their commercial prosperity. The commerce of Glasgow is double that of any other port in Scotland, and is only second to that of Liverpool.

Besides the Clyde there are a few small streams, such as the *Leven*, which drains Loch Lomond, and the *Ayr*, which empties itself into the bay formed by the Firth of Clyde; but these are of little consequence.

(*b.*) The rivers of the Solway Firth consist of four streams thrown down the southern side of the Lowland Hills, the *Dee*, the *Nith*, the *Annan*, and the *Esk*. These rivers flow due south in parallel valleys, and are noted for the beauty of their dales, which are wild and narrow in the upper mountainous districts, but wooded and well cultivated as they broaden in the lower land which borders the coast. The *Dee* traverses in its course the long narrow lake of Loch Ken, and marks the eastern boundary of the Galloway Hills. The *Nith* is remarkable for having its head-waters on the northern slope of the Lowland Hills, so that it travels across the whole breadth of the range in the valley which lies between Black Larg and the Lowther Hills before it opens on its southerly valley of Nithsdale. The *Annan* rises not far from the Tweed, but is thrown down the opposite slope of the mountain range, where it takes up the stream of the *Moffat* as it passes out of the wild dark glen of

Moffat dale. The *Esk*, like the Ettrick, has its source in Ettrick Pen, but flowing southward finally falls into the head of the Solway Firth. Its affluent, the *Liddel*, marks for some distance the boundary between England and Scotland.

Rivers of the Highlands.

Rivers.	Length in Miles.	Area of Basin in Sq. Miles.
Tay	110	2,400
Forth	100	645
Spey	96	1,190
Dee	87	700
Don	50	530

Rivers of the Lowland Hills.

Clyde	98	1,580
Tweed	96	1,870
Nith	60	460

CHAPTER VI.

COAST-LINE AND ISLANDS.

WE have now gone rapidly over the inland features of Scotland : we have seen how its mountain
masses are disposed, where they sink into plains, and
the direction in which they throw down rivers to the
sea. We are therefore better able to understand the
formation of the Scottish coast. For the character
of the shores of any country is determined, not only
by the action of the sea from without, but by the
internal structure of the country itself. Masses of
mountainous rock that present a strong front to the
waves will form a bold coast line with headlands and
peninsulas, just as a low soft shore over which the
sea washes at every tide must gradually be worn away
into deep bays. And broad rivers that flow through
such a low soil to the sea will hollow out great estuaries
and so assist in the slow formation of bays; while
rapid mountain streams as they carve out their channels
among ridges of rock cut the coast into fiords and
sea-lochs, making it jagged and serrated like a saw,
with its teeth of rock projecting into the water.

Form of Coast.—The form of Scotland, with its wild, rugged, and deeply-cut shores, and its outlying fringe of rocky islands, tells the history of the incessant war waged upon the land by the oceans which dash themselves against either coast. At two points, to the north and south, these contending seas have so far advanced into the land as almost to cut the country into two parts, and throughout its whole extent the shore is so deeply indented by arms of the sea that its length is made up to 2,500 miles, or 700 miles more than the coast of England, a country of twice the size of Scotland. By these creeks and bays every part of the land is brought within forty miles of the ocean.

Contrast between the Coasts.—The character, however, of the coast varies greatly on the two sides of the country. Scotland lies, roughly speaking, in the form of a crescent, whose western side curves outward to the Atlantic.   On this western side the coast is bent and rounded, girded with steep mountain ranges jutting out in long lines of rocky peninsulas between which the sea unceasingly dashes and surges, and defended by a double barricade of mountain islands, which lying close to the shore have the appearance of peninsulas, and add greatly to the wild and rugged character of the scenery. For the most part this coast opposes to the ocean a massive wall of rock too hard to be worn into bays, but broken by a series of creeks and sea-lochs, up some of which the waves rush for a distance of thirty miles. These creeks however are not made by the action of the sea ; they are merely the termination of those level valleys and defiles which mark the western watershed of the Highlands, and whose floors, once lifted high above the ocean, have now through the sinking of the whole coast fallen below the level of the water. The only bays of the coast lie on either side of the mountain masses in the south, where the shore is low and yields to the advancing waves.

On the eastern side of Scotland, on the other hand, where the general curve is inwards, so that the coast-line is a good deal shorter than on the west, the shore is mainly formed of soft sandstones and clays, more easily worn away by the action of water ; and though here and there mountainous cliffs have been reared against the sea, yet between these defensive barriers the waves have slowly won their way step by step on the low ground round the river mouths, and bit into the land in wide and deep bays. The peninsulas which bound these bays are large and regular in form ; and the entire absence of islands along the shore forms another point of contrast with the west.

In Scotland, then, there is the same unlikeness between the coasts as in southern Britain, but one far more sternly marked by the fiercer character of its seas and climate. As in England, so in Scotland, we find (1) the wild, deeply-indented mountain wall of the west, and the generally low sandy coast of the east. (2) There is here, too, an island shelter thrown up on the west coast to break the full fury of the Atlantic storms ; the difference being that, while England is shielded by one large island, Scotland has a double row of lesser ones. (3) There is the same general direction of the rivers, by far the greater part being thrown down from the western mountains and emptying themselves into the sea on the east, so as to form at their mouths a number of harbours whose outlook is towards the continent of Europe.

We must now survey these coasts more in detail.

The East Coast, which extends along the North Sea from Duncansby Head to the river Tweed, is divided into two wide curves, whose limits are determined by the mountain-masses of the country.

(*a.*) The northernmost of these curves is enclosed between the main branches of the Highland mountains, which end in the peninsulas of Caithness and Buchan. It receives on its inner side the waters of two river estuaries, the *Dornoch Firth* and the *Moray Firth*, severed by the little peninsula of Ross. The Moray Firth is again divided into two lesser estuaries, the *Cromarty Firth* and the *Firth of Inverness*, which are parted by the Plain of Cromarty. By these Firths the drainage of the northern Highlands is carried to the eastern sea. (*b.*) A second opening in the eastern coast is formed by the inward curve of the land between the mountain masses of the Highlands and those of the Lowland Hills. This curve also narrows to two important estuaries, the Firths of the *Tay* and of the *Forth*, which pour into it the drainage of the southern Highlands, and are parted from one another by the peninsula of Fife.

While the eastern shores are for the most part low and monotonous, they are yet marked by several bold

X

headlands formed by outlying mountain spurs which project into the sea. The chief promontories of the peninsula of Caithness are *Duncansby Head, Noss Head,* and the *Ord of Caithness; Tarbet Ness* lies between the Dornoch and Moray Firths; the peninsula of Buchan ends in the headlands of *Kinnaird's Head* and *Buchan Ness; Fife Ness* severs between the Firths of Tay and Forth; while *St. Abb's Head* marks the extremity of the line of cliffs by which the Lammermuir Hills abut on the North Sea.

The North Coast from Duncansby Head to Cape Wrath fronts the Atlantic Ocean, and is of the same character as the western shore, wild, rugged, cleft with deep fissures, and marked by rocky headlands. The chief among its fiords and promontories are *Dunnet Head* at the mouth of *Dunnet Bay, Whiten Head,* which parts the creeks of the *Kyle of Tongue* and *Loch Eribol,* and *Cape Wrath* by the *Kyle of Durness.*

The West Coast of Scotland, from Cape Wrath to the Solway Firth, is washed by the Atlantic Ocean, and by the North Channel which connects this ocean with the Irish Sea, and which parts Scotland from Ireland at its narrowest point by a distance of but thirteen miles. By the general lie of the land the coast is divided into two nearly equal parts. (*a.*) Its north-western half, from Cape Wrath to Ardnamurchan Point, is marked by the general direction of its sea-lochs, its peninsulas, and its islands, which all trend to the north-west. The inlets of the sea are for the most part narrow openings between the cliffs with steep promontories marking their entrances, and lie ranged closely in parallel lines throughout the whole length of the coast. (*b.*) Very distinct from this is the character of the south-western half of the shore from Ardnamurchan Point to the Solway Firth. Here the promontories, sea-lochs, and bays are all ranged in parallel lines bearing to the south-west. There is a change too in the character of the coast: the openings take the form of bays, and are larger and wider, with lesser branches which run far inland among the mountains as sea-lochs. Being larger, these bays are also fewer in number, and are bounded by more important and well-marked peninsulas.

The principal *bays* of the western coast are *Loch Edderachillis*, famous for its herring fisheries; *Loch Enard;* the double inlet of *Loch Broom*, with a good anchorage for boats; *Loch Torridon*, one of the finest of Highland bays; *Loch Carron*, with its carefully cultivated shores; the safe harbour of *Loch Alsh ;* the fissure cleft by *Loch Hourn* through steep wooded rocks; and *Loch Sunart*, opening to the Sound of Mull.

To the south of these the coast is cleft by three great inlets of the sea with openings to the southward, the *Firth of Lorn*, the *Firth of Clyde*, and the *Solway Firth*. (1.) The Firth of Lorn, which is sheltered to westward by the island of Mull, lies at the south-western opening of Glenmore, and thus corresponds to the Moray Firth of the eastern coast; it narrows into a second inlet, *Loch Linnhe*, which again sends out several lesser branches, *Loch Etive*, *Loch Leven*, and *Loch Eil*. (2.) The Firth of Clyde is formed by an arm of the sea

which encroaches on the Lowland Plain between the
Highlands and the Lowland Hills, and thus answers
to the Firth of Forth on the opposite coast. This bay,
which is almost landlocked by the long peninsula of
Cantyre, narrows to the estuary of the river Clyde,
and further on forms two sea-lochs, *Loch Long* and
Loch Fyne, which stretch into the Highland country in
a northerly direction. (3.) The Solway Firth cuts off
the Hills of Galloway from the Cumbrian mountains
of England. It lies at the mouths of several streams
thrown down from the surrounding heights, and is of
little commercial importance. The northern shore
of the Firth is broken by *Luce Bay* and *Wigton Bay.*

The chief *headlands* of this coast lie in the Highland
district; among them are *Cape Wrath, Ardnamurchan
Point,* and the *Mull of Cantyre.* This last promontory
forms the extremity of a very striking peninsula, that
of *Knapdale and Cantyre,* which extends southwards
in a tongue of land forty miles long, and varies from
half a mile to eight miles in breadth till it reaches to
within thirteen miles of the Irish coast. The chief
headland formed by the Lowland Hills is the *Mull
of Galloway,* which lies at the extremity of the peninsula called the *Rhinns of Galloway,* once an island,
but now connected with the shore by a narrow neck
of land.

The Islands.—The Scotch islands lie wholly to
the north and west, where they form a huge double
breakwater against the violence of the Atlantic storms,
the northern groups consisting of the *Orkneys* and the
Shetlands ; the western groups of the *Inner* and the

Outer Hebrides. Besides these groups of islands there is another cluster lying within the Firth of Clyde. The whole number of islands amounts to 788, of which however the greater part are bare rocks and crags ; in fact, only 186 of them are inhabited, while some of these contain but two or three people.

I. The *Orkney Islands*, lying in the north-east, are separated from Scotland by the wild tempest-driven strait of the *Pentland Firth*, only six miles in width. They contain sixty-seven islands, many not inhabited : the islands are generally low and flat, broken by rounded hills, and having steep rugged cliffs on the west, those of *Hoy* being the finest sea-cliffs in all Scotland.

II. The *Shetland Islands*, more than 100 in number, lie far away in the northern seas, fifty-six miles beyond the Orkneys. For the most part they are mere heathy wastes broken by rocks, and fenced on both sides by gigantic cliffs worn into wild forms by the violence of the Atlantic tempests. More than half of them have no inhabitants. The climate is very wet, but the cold is not severe, and good corn and grass can be grown in their sheltered plains.

III. The *Inner Hebrides* lie close to the western coast; they are mainly composed of masses of volcanic rock. To the northward rises the large irregular *Isle of Skye* with the splendid group of the *Coilin Mountains* towering over its red dome-shaped hills, parted from the shore by the narrow *Sound of Sleat.* From thence a line of scattered islets leads to the island of *Mull*, which is cut off from the mainland by the

narrow *Sound of Mull;* here the mountains are green
and terraced, with wide stretches of heathy uplands
between, while above all towers the great mass of *Ben
More.* Some way to the south of Mull is *Jura*, with
its group of lofty cones of glistening quartz, pale
gray and white, the *Paps of Jura ;* and parted from
it by the narrow *Sound of Islay*, the island of *Islay*,
broken by low hills, with cultivated fields between
them, the most fertile spot of the Hebrides. *Colonsay*
and *Oronsay* are of very small extent, and form at low
water but one island.

IV. To westward of this inner group of isles lies a
second breakwater formed by the *Outer Hebrides*, a
long chain of islands ranged in close order, but torn
and broken by the incessant beating of the waves.
They are parted from the inner Hebrides by the straits
of the *Minch* and the *Little Minch.* Their chief islands
are *Lewis* and *Harris*, united by a narrow neck of land,
North Uist, *Benbecula*, *South Uist*, and the *Barra*
islands. Lewis is, for the most part, moss and moorland ; Harris and most of the other islands are rocky
and broken by steep mountains.

V. The last group of the Scotch islands are enclosed
in the Firth of Clyde, land-locked and sheltered to
westward by the rocks of Cantyre. Close to this
peninsula lies *Arran*, the largest among them, whose
southern half is formed of basalt and other rocks such as
those which break the Lowland Plain, and whose chief
summit, *Goat Fell*, is the highest mountain of southern
Scotland. To the north of it lies *Bute*, with its low
wooded hills, only severed from the mainland by a

narrow strait, the *Kyles of Bute.* In contrast to these larger isles, the two little islands called the *Cumbrays* are green, level and bare, forming the links of a natural bar drawn along the mouth of the Firth.

Bays.

North Coast.	East Coast.	West Coast.
Dunnet Bay.	Moray Firth.	Edderachylis Bay.
Kyle of Tongue.	Dornoch Firth.	Enard Bay.
Loch Eribol.	Cromarty Firth.	Loch Broom.
Kyle of Durness.	Inverness Firth.	Loch Torridon.
	Firth of Tay.	Loch Carron.
	Firth of Forth.	Loch Alsh.
		Loch Sunart.
		Firth of Lorn.
		Loch Linnhe.
		Firth of Clyde.
		Loch Fyne.
		Loch Ryan.
		Solway Firth.
		Luce Bay.
		Wigton Bay.

Headlands.

Duncansby Head.	Ord of Caithness.	Ardnamurchan Pt.
Dunnet Head.	Tarbet Ness.	Mull of Cantire.
Whiten Head.	Kinnaird Head.	Mull of Galloway.
Cape Wrath.	Buchan Ness.	
	Fife Ness.	
	St. Abb's Head.	

Islands of the West Coast.

Outer Hebrides—   Area in square miles.

    Lewis and Harris 748

    South Uist 127

    North Uist 118

    Benbecula 43

Inner Hebrides—

    Skye 535

    Mull 330

    Jura 130

    Islay 240

Islands of the Firth of Clyde—

    Arran 220

    Bute 65

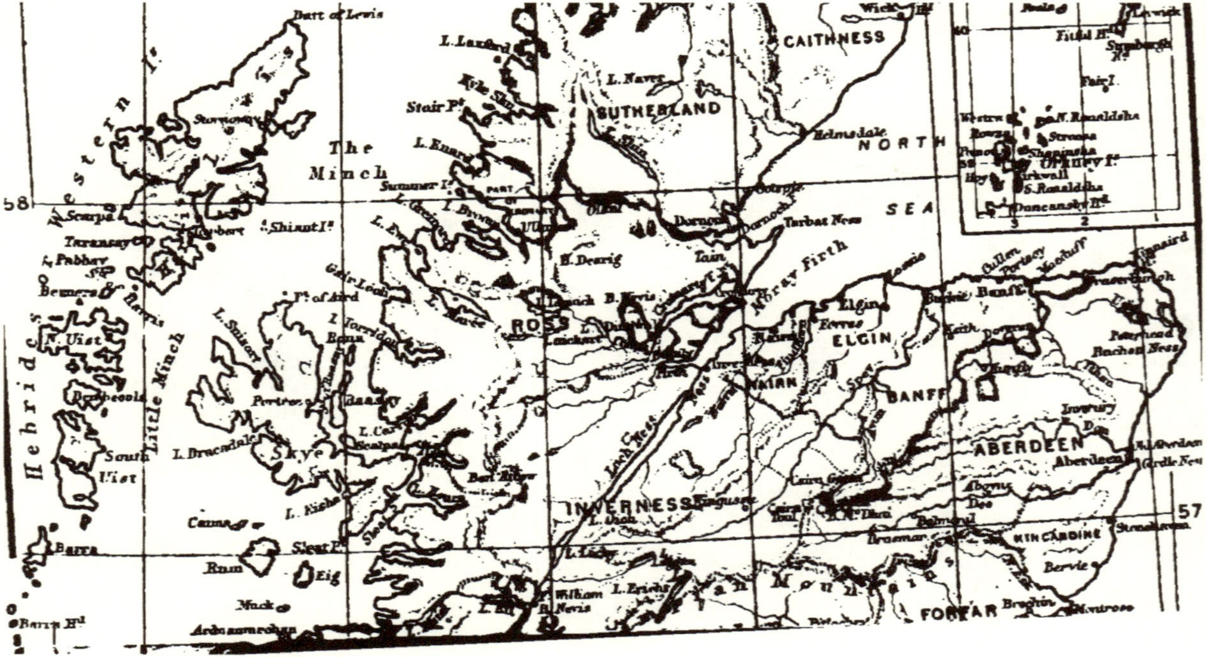
Butt of Lewis
L. Laxford
Stair Pt
L. Enard
The Minch
Summer Is
Shiant Is
Scarpa
Taransay
Pabbay
Harris
Berneray
N. Uist
South Uist
Little Minch
Hebrides
Western Is
Barra
Barra Hd
L. Snizort
Portree
Skye
Raasay
L. Dracadale
Canna
Rum
Eig
Muck
Ardnamurchan
Sleat Pt
L. Torridon
Gair Loch
Pt of Aird
L. Maree
L. Broom
L. Carron
L. Naver
SUTHERLAND
CAITHNESS
Wick
Helmsdale
NORTH SEA
Dornoch
Tarbat Ness
Tain
Moray Firth
ROSS
B. Dearig
B. Wyvis
INVERNESS
Loch Ness
L. Lochy
L. Ericht
William
B. Nevis
NAIRN
ELGIN
Elgin
Forres
BANFF
Banff
Cullen
Portsoy
Macduff
Kinnaird Hd
Fraserburgh
Peterhead
Buchan Ness
Keith
Inverury
Don
ABERDEEN
Aberdeen
Girdle Ness
Aboyne
Dee
Balmoral
Braemar
KINCARDINE
Stonehaven
Bervie
Brechin
Montrose
FORFAR
Cairn Gorm
Fair I.
Orkney Is
Kirkwall
Stromness
Hoy
S. Ronaldsha
N. Ronaldsha
Westray
Shapinsha
Duncansby Hd
Foula
Fitful Hd
Sumburgh Hd
Lerwick
58
57

SCOTLAND,

English Miles

ATLANTIC OCEAN

NORTH CHANNEL

ENGLAND

FIFE

BERWICK

ROXBURGH

DUMFRIES

AYR

WIGTON

ARGYLL

Atlantic

W. & A. K. Johnston Edinburgh and London

Longitude West 6 of Greenwich 5

CHAPTER VII.

SCOTLAND, like England, is divided into a number of shires or counties. But here this division is of comparatively modern times, and it so happens that some of the early territorial names of different parts of the country which existed long before any shires had been formed have been preserved, and are still largely used to describe districts which do not in most cases coincide with any county boundaries. We shall briefly notice the more important among these ancient divisions, before passing on to the counties themselves.

Local Divisions of the Highlands.—In our study of the Highland mountains we have already been brought across most of their local divisions. Along the western coast we have noted the jutting peninsulas of *Glenelg, Morven, Appin, Lorn, Knapdale and Cantyre*, and *Cowal*. On the borders of the North Sea too we have described the hilly country of *Ardross* and the low fertile plains of *Moray* bordering the Moray

Firth, with the valley of *Strathspey*, that parts them from the hills of *Buchan*. These in their turn have to southward of them the district of *Mar*, which occupies the middle part of the valley of the Dee; from whence we pass into the Lowland Plain across *Mearns*, or the low ground between Mount Battock and the sea-coast, and by *Angus*, where the eastern Highlands abut on the valley of Strathmore in the long spurs known as the Braes of Angus.

The more important districts which occupy the central part of the Highlands are *Lochaber* and *Badenoch*, situated on either side of the waterparting as it crosses Glenmore. Lochaber is marked by the mass of Ben Nevis in its centre, while Badenoch comprises the wild country that lies from Glenmore to the Cairngorm Mountains, and is traversed by the upper valley of the Spey. Immediately to the south of Badenoch Forest lies the equally wild and mountainous district of *Athol*, stretching down to the gorge of the Garry; and beyond this valley rise the lofty summits and ridges of the *Breadalbane* territory which terminate on the Lowland Plain, and from the midst of which open valleys such as those of *Strathearn* and *Menteith*.

Local Divisions of the Highland Plain.— We are thus brought directly into the Lowland Plain. In this we have already traced the limits of *Strathmore;* a part of this valley, which extends across the river Isla, is known as *Stormont*, and the narrow belt of land shut in between the Sidlaw Hills and the Firth of Tay as the *Carse of Gowrie*. A broad tract of fertile soil which forms the southern border of the Firth of Forth constitutes the district of the *Lothians*, divided into *East, Middle,* and *West Lothian.* The western side of the Lowland Plain has also its territorial divisions. *Lennox* lies to the south and south-east of Loch Lomond. *Cunninghame* and *Kyle* occupy the larger part of the plain of Ayrshire.

Local Divisions of the Lowland Hills.— The most familiar among the names of all these ancient districts of Scotland are those which we find in the Lowland Hills, and which are in as common use as the names of the shires themselves. The south-western peninsula formed by these hills is divided into two well-marked portions—*Carrick*, which occupies the north-western watershed towards the North Channel; and *Galloway*, or the south-western watershed from Luce Bay almost to the valley of the Nith.

The central part of the Lowland Hills is traversed by a number of river dales, each of which has given its name to a separate district of well-marked geographical character : *Clydesdale* lies to the northward ; to the southward, on the other hand, are *Nithsdale, Annandale, Eskdale,* and *Liddesdale.*

Other river-dales, as we have seen, penetrate into the heart of the Lowland Hills in the north-east. The upper valley of the Tweed forms *Tweeddale*, while into its lower valley opens the dependent districts of *Ettrick Forest, Teviotdale,* and *Lauderdale ;* the place where the river pours itself into the sea, and where Berwick now stands, was in old times known as the *Merse.*

The Scotch Counties.—The ancient divisions

of which we have just spoken have however no relation to the civil government of Scotland, and for all political purposes the country is divided into *thirty-three* shires or counties.

In geographical character these counties differ from those of England just as the two countries differ in physical structure. We have seen that in England the river-system of the land has so impressed itself on the political divisions of the country as to form a most important element in determining the position and grouping of the shires. The boundaries of these shires and groups of shires are not mere artificial lines, but landmarks clearly defined by the natural forces which have set limits to the river-valleys and river basins in which they lie. But in Scotland, where the river-system is of a wholly different character, this is not always the case. It is true that in the Lowland Hills and in the Lowland Plain, where we find the running waters drawing together into river-basins, the shires do in fact group themselves, as in England, into knots of counties whose areas coincide with that of the river-basin in which they lie, and whose limits are determined with exactness by the local waterpartings that divide river from river. But in the district of the Highlands, with its multitude of independent mountain torrents, such an arrangement of counties is impossible. At the same time the river-system even here plays a great and important part in their formation.

In the southern Highland districts, where the

country consists half of mountain and half of plain, the limits of the counties still mainly coincide with those of the river-basins. As these basins, however, shrink farther northward into mere straths or glens, the counties spread over the land between them and the rivers themselves thus become the shire boundaries. And when still more to the north the streams become mere mountain torrents of insignificant size, the shires spread over the waterparting ridge itself, and extend down its watershed to east and west.

In our study of the Scotch counties, therefore, we must never lose sight of the determining influence exercised by the river-system of the land, and as in England so in Scotland, we shall take the line of the waterparting and the lie of the greater river-basins to guide us in grouping the shires together. Owing however to the character of the country and the strongly-marked features which distinguish its three main divisions from one another, it will be convenient to divide the Scotch counties into three groups which answer to these parts, and which at the same time follow the course of the waterparting line. These groups are (1) the counties of the Highlands; (2) the counties of the Lowland Plain; (3) the counties of the Lowland Hills.

CHAPTER VIII.

THE HIGHLAND COUNTIES.

THIS important cluster of shires covers nearly two-thirds of Scotland, stretching as it does from the Shetland Isles to the Firths of the Tay and of the Clyde, and occupying a district of 18,600 square miles in extent. It consists of fourteen distinct counties, which are naturally divided into two groups both by geographical position and by a certain difference in their physical characteristics. These are (1) *The Counties of the Northern and Western coasts;* and (2) *The Counties of the Eastern coast.* Each group contains seven shires.

[A] THE COUNTIES OF THE NORTH AND WEST HIGHLANDS, which are all washed by the Atlantic, comprise (1) The Orkney and Shetland Islands, (2) Caithness, (3) Sutherland, (4) Ross and Cromarty, (5) Inverness, (6) Argyle, (7) Bute. All these shires, save those that are wholly insular, are linked together in a very marked way by the main waterparting of the Highlands, which passes through each of them in turn.

They are distinguished too by the same general features, extending as they do over a district which, as we have seen, is wholly built up of a succession of high mountains, barren moors and mosses, forest-land, and deep valleys or straths, with lakes embedded in them. Till of late the country has been thought unfit for agriculture from the harshness of its climate and poverty of its soil, and formed a vast stretch of pasture-land for a hardy race of sheep and cattle Owing to its scanty resources, save that of sheep-grazing, the people inhabiting it are exceedingly few : there is not a single town throughout the whole central part of the district, and very few hamlets ; only along the sea-coast runs a thin fringe of little sea-ports supported by fisheries of herring, cod, and salmon, and by the trade of exporting these fish to the Continent and to England.

I. **The Orkney and Shetland Islands** form together one county, whose chief town, *Kirkwall*, lies in *Pomona*, or *Mainland*, the largest island in the Orkneys. The majority of the Orkney islands are low and fertile, and broken only by rounded hills whose sides are covered with grass : twenty-seven of them are inhabited, while about forty more serve as pasture-lands for cattle. Of the 100 islands which make up the group of the Shetlands, but thirty-four are inhabited. The rest form mere *stacks* and *skerries* of rock, the first rising sharply like chimneys from the sea, the second consisting of low broken reefs. The chief island of the Shetlands is also known as *Mainland*, and has in it the little town of *Lerwick*.

II. The shire of **Caithness** forms the north-eastern extremity of Britain, and has but one land boundary, that on its western side, where a ridge of hills running north from Morven and the Ord of Caithness parts it from Sutherland. Through its centre passes the north-eastern extremity of the waterparting that divides between the watersheds of the Atlantic and the North Sea; but the shire, consisting as it does throughout its whole extent of Old Red Sandstone, is as sharply marked off from the mass of the Highland mountains by the structure of its rocks as by its peninsular form; and this difference is strikingly represented in the comparatively low and level character of the ground and the fertility of the soil. The bulk of the county is made up of monotonous moorland surrounded by rugged and sea-worn cliffs, from whose heights the ancient keeps of the northern vikings looked out over the Atlantic. The bleak moors and marshes of the interior are gradually being cultivated by the industry of the people, and now form one of the most fruitful districts of northern Scotland. The chief town, *Wick*, and a second village, *Thurso*, are built on the sea-coast, and are chiefly supported by herring fishery. The population amounts to 40,000 people.

III. The red sandstone hills of Caithness give place westward to the mountain masses of Silurian rock which compose **Sutherland.** This shire stretches across Scotland from sea to sea: to northward it is washed by the Atlantic, and to southward it is parted from Ross by a line drawn from the

Dornoch Firth to Enard Bay. The rough square thus formed has an area of 1,886 square miles, and consists of a continuous stretch of naked and barren moorland broken by masses of mountains, such as those which mark the water-parting, Ben Hope, Ben Klibrek, and Ben More ; while beyond these again to westward a fringe of ancient gneiss borders the Atlantic with a line of cliffs deeply cut and indented by creeks and inlets of the sea. Since the high ground thus lies in the western half of the shire, the bulk of its running waters are poured eastward through long valleys into the Dornoch Firth, the chief of these glens being marked by lakes such as Loch Shin, which with its series of dependent lochs nearly cuts the county in two parts. From the character of the soil Sutherland remains the bleakest, wildest, least cultivated, and least populated county in the whole of Britain : it contains, throughout its entire extent, about 24,000 inhabitants, and its county town, *Dornoch*, is a mere fishing village on the Dornoch Firth.

IV. **Ross** and **Cromarty** form practically but one county, Cromarty being made up of a number of small parts scattered throughout Ross-shire. The united shires form a triangular tract of country of nearly 4,000 square miles in extent, which stretches across Scotland from sea to sea, with its apex pointing eastward, and with its broad end resting on the Atlantic. Its northern boundary from Loch Enard to the Dornoch Firth touches Sutherland, while on the south, from Loch Beauley to Ben Attow, it marches with Inverness.

Y

(*a.*) The broad western part of the shire is traversed by the waterparting range as it runs from
Ben More to Ben Attow, and throws out to the
westward those lateral ridges of rock which border
the Minch and the Inner Sound in a series of rugged
peninsulas parted by deep bays. The chief among
these bays are Loch Ewe, which almost forms a
continuation of the inland lake, Loch Maree, and
further south, Loch Torridon and Loch Carron.

(*b.*) The eastern half of the united shire consists
mainly of the mass of Ben Wyvis, the lower hills of
Ardross, and the flat plain of Cromarty, a spit of
fertile soil composed of Old Red Sandstone which projects between the Dornoch and Moray Firths, and so
forms the apex of the triangle. Into these two firths
is poured nearly the whole drainage of the county by
the gathering together of the mountain torrents thrown
down to the eastward from the waterparting range.
The river valleys, however, are mere glens, and the bulk
of the shire remains a continuous mass of barren moorlands and mosses, containing but a small population,
about 80,000 in all. The strip of fertile soil which
borders the eastern shore and encloses the estuaries
of the Dornoch Firth and Loch Beauley is alone
fitted for habitation; and here lies a group of little
towns scarcely larger than villages, such as *Dingwall*,
the county town of Ross and Cromarty; *Tain*, in Ross;
and *Cromarty*, at the mouth of the Cromarty Firth.

The island of Lewis also belongs to the county of
Ross; it is 557 square miles in extent, mountainous
in parts, but with large tracts of bog and moorland.

Its chief town, *Stornoway*, lies on a little harbour on its eastern coast.

V. To the south of Ross stretches the largest shire in Scotland, that of **Inverness**, which comprises an area of 4,255 square miles. It may be best described as consisting of the country which lies on either side of the cleft of Glenmore, and which opens to the eastern sea by the Moray Firth, and to the western sea by the shore of Sleat Sound. With the adjoining island of *Skye*, and the bulk of the *Outer Hebrides*, which form part of its territory, the whole shire has somewhat of the form of a vast crescent; to the north it half encircles Ross, while its southern boundary, passing by Loch Shiel, Loch Eil, and Loch Ericht, is continued along the Badenoch and Cairngorm Mountains, and then turns northward across Strathspey to the Firth of Inverness. The shire thus borders with Argyle and Perth to the southward, and to eastward with Aberdeen, Elgin, and Nairn.

Like the northern counties Inverness consists of two well-marked divisions. (*a*.) To westward lies the high ground of the waterparting from Ben Attow to the Moor of Rannoch. Its dependent ridges of rock thrown out to the coast are defined by their enclosed lateral glens and lakes, such as Loch Morar and Loch Shiel dipping to the Atlantic, and Loch Arkaig falling toward Glenmore; while to the south of the Great Glen the mass of Ben Nevis rises from the moorlands of the Lochaber district. (*b*.) The eastern half of the county, on the other hand, is made up of alternate chains of hills and river valleys, through which the

drainage of the mountains is poured into the Moray
Firth. To the north lie Strathglass and Glengarry;
these are parted from Glenmore and Loch Ness by a
belt of hills of which Mealfourvounie is the most important; Glenmore in its turn is severed from Strathspey by the Monadh Leadh Mountains; while Strathspey is bounded to the south by the Badenoch and
Cairngorm heights.

Inverness thus consists of a vast mountain tract
only broken by wild glens and straths wholly unfitted
for habitation or agriculture. It contains indeed one
town, *Inverness*, which lies on the Firth of the same
name, in the midst of the narrow fringe of fertile
sandstone that borders the eastern shore, and which
with its 14,000 inhabitants forms the capital of the
Highland district. The opposite extremity of Glenmore is guarded by a military post, *Fort William*.
But the broad reaches of moor and mountain which
constitute the interior of Inverness do not contain
a single town; and among the hamlets scattered
throughout them the village of *Kingussie* on the
upper Spey alone derives some consequence from
the importance of its position at the opening of the
Pass of Dalwhinnie, which leads from Strathspey to
the valley of the Tay. With its scanty population of
87,480 persons scattered over this broad expanse of
territory, Inverness forms next to Sutherland the most
thinly-peopled of Scotch shires.

The Island of Skye, which is included in the shire
of Inverness, is the largest of the Inner Hebrides,
having an area of 535 square miles. Its chief town

is the little fishing village of *Portree*, sheltered under the rugged rocks of which the island consists, and which in the south tower up into the heights of the Cuchullin Hills, 3,000 feet above the sea. Besides Skye, Inverness also includes within its shire-limits most of the islands of the Outer Hebrides—Harris, North and South Uist, Benbecula, and the Barra Islands.

VI. The line of water-parting as it passes southward from Inverness strikes along the border of **Argyleshire**, where it forms a boundary that divides this county from Perthshire to the east. Argyleshire thus occupies that part of the western watershed of the Highlands which stretches from Ardnamurchan Point to the Firth of Clyde, and with its area of 3,255 square miles forms a county of very great extent. Lying as it does along the wildest and most fantastic part of the western coast, where innumerable inlets of the sea cut deep into the land, the county is broken into a series of peninsulas all ragged and torn, such as those of Morven, Appin, Lorn, Cowal, and the long tongue of land formed by Knapdale and Cantyre as it stretches southwards towards the Irish coast. The shire includes too a number of islands which border the coast and which, though they once formed part of the land, have been severed from it by the gradual sinking of the shore, and have been left standing in mid-sea. The largest of these islands, Mull, Jura, and Islay, and the lesser ones of Iona and Staffa, almost belong to the mainland: Coll, Tiree, and Colonsay lie somewhat more apart.

The interior of Argyleshire consists of the south-
western offshoots of the waterparting range as they
are thrown westward to the sea between the deep
fiords of Loch Lynnhe with its several branches, Loch
Fyne, and Loch Long, and an inland lake, which in
form resembles these fiords, Loch Awe. The mountains
rise to their greatest height at the centre of the county
in Ben Cruachan, whence they gradually sink downwards towards the ocean. But in the islands the
mountain peaks again rise in grandeur and wildness,
as in the great mass of Ben More in Mull, and the
precipitous Paps of Jura, and form, as it were, massive
bulwarks to protect the inner coast-line. Herring-
fishery and sheep-grazing support a small population of 75,680 people, who are gathered along the
sea-coast in a thin fringe of insignificant towns. The
little county town of *Inverary* lies on the shores of
Loch Fyne; *Campbelltown*, on the Cantyre peninsula, is a place of more importance; *Oban*, on the
southern shore of Loch Lynnhe, above the mouth of
the Caledonian Canal, is famous for its picturesque
scenery; *Dunoon* is a watering-place which lies on the
Firth of Clyde.

VII. The shire of **Bute** is made up of the islands
of the Firth of Clyde, shut in between the peninsula of
Cantyre and the mainland. These are *Arran, Bute,
Great* and *Little Cumbray*, and *Ailsa Crag.* The
largest of these, Arran, measures twenty miles in length
and eleven in breadth, while Bute is only fifteen miles
long and five broad. Both islands belong by their structure half to the Highlands and half to the Lowlands.

Their northern part is formed of the granite, mica, and clay slate of the Highland mountains, which rise into lofty hills such as Goat Fell, 2,874 feet, in Arran; while in the southern part of the islands appear the red sandstone, carboniferous limestone, and igneous rocks of the Lowland Plain. The county town is *Rothesay* in Bute. In an area of 371 square miles, this shire maintains nearly 17,000 inhabitants.

[B.] THE COUNTIES OF THE EASTERN HIGHLANDS, which adjoin the North Sea, are also seven in number ; but these cover a much smaller tract of land than those that lie along the course of the waterparting. They · are (1) Nairn, (2) Elgin, (3) Banff, (4) Aberdeen, (5) Kincardine, (6) Forfar, and (7) Perth. Their position within the Highland district gives them a wild and mountainous character which resembles that of the western counties. But lying as they do along the eastern shore and extending southward over the fertile plain of Strathmore, they embrace all the larger straths and river-basins of the Highlands and the flats by which the sea-coast is bordered. They thus possess in parts a considerable agricultural importance, especially in Buchan and the valley of the Dee, districts famous for their cattle ; and they have besides an important manufacturing and trading industry which is carried on in the district along the shores of the Firth of Tay. These various means of wealth place the eastern counties in a very different position with regard to population from the shires of the north and west, for though they do not cover one half the extent of land, they contain more than double the number of inhabitants.

The range of mountains which extends from Dalna-. spidal to Stonehaven runs like a wall across the country occupied by this group of shires, and from its un- broken length and great height has since the earliest times of Scotch history formed a most important boundary between the tribes and peoples of this dis- trict. It marks, roughly speaking, a broad difference in geographical character between the shires which lie to northward of it and those which are situated to the south, a difference that results from the character of the river-systems of these two districts. The moun- tain torrents to the northward form narrow straths running in parallel lines, and the shires lie extended for the most part across the tracts of hilly country that sever between these straths. The southern shires, on the other hand, occupy the basins of larger rivers, rivers which flow through the midst of the counties, and in so doing gather in lesser streams and unite into river-systems.

I. The low land that borders the Moray Firth, and part of which was formerly known by the name of *Moray*, has been broken up into a number of very small shires, which occupy the tracts of land enclosed between the river-valleys. The first of these little counties, Nairn, is made up of a belt of hilly country, 215 square miles in extent, which lies between the streams of the Nairn and the Findhorn, and is traversed by the last spurs of the Inverness mountains as they fall to the strip of Old Red Sandstone which borders

the sea-shore.   This fringe of low land is capable of
tillage, and by means of a little agriculture and by
the fisheries of the coast the shire is able to support about 10,000 people.   Its county town, *Nairn*,
lies on the harbour formed by the river of the same
name.

II. To eastward of Nairn lies a second small shire,
Elgin, which is still sometimes known under the old
name of this district, **Moray**.   The last spurs of the
Monadh Leadh Mountains, which strike through its
centre from Inverness, fall westward to the valley of
the Findhorn, and eastward to that of the Spey, while
towards the sea-coast they give place to a broad tract
of Old Red Sandstone, where the soil is fruitful and
well cultivated.   With an area of 530 square miles,
Elgin is twice as large as Nairn, and contains a population more than four times as great, having over
43,000 inhabitants.   It possesses two small towns—
the capital, *Elgin*, which lies on the little stream of
the Lossie, and the more ancient town of *Forres*, at
the mouth of the Findhorn.

III. The Old Red Sandstone rocks which play so
large a part in Nairn and Elgin stop short at the
valley of the Spey, and do not cross into the shire
of **Banff**, which lies to westward of it.   This county
consists of a long and narrow wedge of hills of Silurian rock driven in between the valleys of the Spey
and the Doveran.   Beginning on the south-western
border of the shire in the huge mass of Ben Macdhui,
the chain of heights is prolonged by Ben Rinnes and
Corryhabbie, and only dies down into low hills as it

nears the coast. The whole extent of Banffshire is 686 square miles ; its fisheries and agricultural industries maintain a population of about 62,000 persons, chiefly gathered round the little harbours of the seacoast. The county town, *Banff*, lies at the mouth of the Doveran.

IV. A far more important county than these last, both for size and population, is the shire of **Aberdeen**. Extending from the mouth of the Doveran to that of the Dee, it forms in the peninsula of Buchan, which projects into the North Sea, the easternmost part of Scotland, while to the westward it stretches along the valleys of the Don and of the Dee into the heart of the Highlands, to the point where the mountains break into two distinct lines of high ground at the source of the Dee. On one side of this valley Ben Avon, Ben Macdhui, and Cairntoul part it from the shires of Banff and Inverness; while on the other side, Mount Keen, Ben-y-Gloe, Loch-na-gar, and Mount Battock, sever it from the counties of Perth, Forfar, and Kincardine.

The shire of Aberdeen, with its area of 1,970 square miles, thus consists of two parts. The peninsula of Buchan is occupied by that belt of high ground which parts between the rivers of the Moray Firth and those of the North Sea, and which here dies down into a tract of low hilly country that forms the best grazing land in Scotland. From it come the finest cattle with which the London market is supplied. This agricultural district contains no towns save a few small seaports, such as *Peterhead* and *Fraserburgh*, which are engaged

in the fisheries of the coast, and in the whale and seal
fishery of the northern seas.

On the other hand, the bulk of the county to the
southward consists of the valleys of the Dee and the
Don, with the masses of hills which part them from
each other. These valleys form, in fact, wild mountain
straths famous for the grandeur of their scenery, but
containing merely a few little hamlets : *Braemar* on
the Dee is much visited by tourists, and below it lies
the palace of *Balmoral*. But as the Dee pours itself
into the sea, it forms a harbour on whose shores has
grown up the town of *Aberdeen*, one of the chief trading
cities of eastern Scotland. Situated on the sea-coast, at
the opening of two important valleys, and commanding
the road to northern Scotland, Aberdeen became very
early a city of much importance, carrying on a large
trade with northern Europe. It now adds to its manufactures of cotton, silk, and wool, the construction of
iron steam-vessels, and the working of vast quarries of
slate and granite in the neighbouring country ; while
besides these industries, it forms the chief centre of
export for the great cattle trade of the north. Aberdeen
has thus grown into the fourth greatest city of Scotland,
and with its 88,000 inhabitants has gathered into
itself more than a third part of the population of the
shire in which it lies, whose total number amounts
to over 244,000.

V. The three remaining shires of this group,
Kincardine, Forfar, and Perthshire, belong only in part
to the Highlands, for occupying the southern limits of
this great mountain mass, they extend into the valley

of Strathmore, and thus emprace a part of the Lowland Plain.

The smallest of these shires, that of **Kincardine,** formerly known as the *Mearns,* consists of a little tract of land which lies along the shore between the valleys of the Dee and the Esk, and extends westward to Mount Battock ; it thus forms the great route which leads into northern Scotland between the mountains and the sea. The slate rocks which occupy the shire on the Aberdeen border give way in the south to a belt of Old Red Sandstone which constitutes in fact a part of the valley of Strathmore as it. extends into the shire from Forfar. The north-eastern opening of this valley is marked by the little county town, *Stonehaven,* which lies on the sea-coast and carries on a small fishing trade.

The shire, only 383 square miles in extent, contains about 34,000 inhabitants.

VI.   Both the slate mountains and the red sandstone plain of which the shire of Kincardine is made up are continued to the south-west into the county of **Forfar,** which coincides with the older district of *Angus.* Its northern border adjoining Aberdeen is formed by a chain of heights that runs from Mount Battock to Glas-Mhiel, and the southern slopes of this range as they pass into Forfar are known as the Braes of Angus. Along their base on the south-east stretches the level plain of Strathmore, beyond which, by the shore of the Firth of Tay, rises the lower chain of the Sidlaw Hills. The three parallel belts of country thus formed are all alike prolonged across the

south-western border of the shire and are carried
thence into the adjoining county of Perth.

In Forfar we have almost passed out of the Highland district, for the great bulk of the shire is occupied
by the Lowland valley of Strathmore, a plain watered
by streams thrown down from the northern heights, as
the North and South Esk, and rich in corn and grass-
lands. In its centre lies the county town, *Forfar*,
with 11,000 inhabitants, a town little smaller than
Inverness, though that is the capital of a shire 1,100
square miles greater in mere extent than Forfar. But
the main wealth of the shire does not lie in its agricultural industries, but in an important trade which has
been developed in a group of sea-port towns whose
position on the coast, near the coal-fields of the central
plain, and with easy access to them by water, gives them
the double advantage of manufacturing towns and ports
for foreign trade. The greatest of these towns, *Dundee*,
lies on the Firth of Tay ; to its immense manufactories
of jute, coarse linen, sail-cloth, and sacking, it adds a
very large shipping trade, and with its population of
120,000 people, ranks third among the great towns
of Scotland. To north of it lie the ports of *Arbroath*
and of *Montrose* on the South Esk, both of which
have large linen manufactures. At a little distance
from Montrose is the lesser town of *Brechin* on the
Esk.

The manufacturing and commercial industries of
Forfar have made it one of the most populous
shires in Scotland. In its area of 875 square miles it
contains 237,528 inhabitants.

VII.   If we now pass westward out of Forfar we
enter on a third county which occupies the border-land
between Highlands and Lowlands, the county of **Perth.**
This shire is about equally divided between the mountains and the plain.   To the north the chain of heights
which runs from Glas-Mhiel to Loch Ericht and parts
it from Aberdeen, throws out southward a mass of
mountains and hill-ridges which forms the district of
Athol.   To the west lies the water-parting range
from the Moor of Rannoch to Loch Lomond, which
divides Perthshire from Argyle, and which, by the
lateral ridges of rock that it sends out to the eastward,
constitutes the group of the Breadalbane Mountains.
By these hill-masses Perthshire forms a part of the
Highland district.   But along their base to the south-
east strikes the red sandstone valley of Strathmore,
which connects it with the Lowland Plain.   This
valley, as we have seen, is continued from Forfar in
the north-east across the Isla into Perth, and is
thence prolonged across the Forth into Stirling; while
it is shut in to the south-west by the igneous rocks
of the Ochill and Sidlaw Hills, which part Perthshire
from Fife, Kinross, and Clackmannan.

Perthshire, made up of these two divisions, forms
a vast irregular circle with an area of 2,834 square
miles, which almost exactly coincides with the basin
of the Tay.   This river, rising in the Breadalbane Mountains, gathers in the drainage of the
western heights brought to it by such streams as the
Earn and the Tummel, while from the northern
mountains are poured down the Garry and the Isla.

All these rivers as they traverse the plain unite their waters into one system and empty themselves into the sea through the Firth of Tay. Part of a second river-basin lies along the southern border of the county, where the Forth and its affluent, the Teith, break in like manner from the mountains and flow across the sandstone plain at their base.

Perthshire therefore occupies a geographical position of the first importance. The only one of the Highland counties which lies wholly inland, it is placed in the very heart of Scotland, and includes within its limits not only the bulk of the rich fields of Strathmore, but also every one of the main passes that lead from the Lowland Plain into the centre of the Highlands.

The mountainous districts of Perthshire really consist, as we have seen, of a series of abrupt lateral ridges by which the Highlands suddenly terminate on the Lowland Plain. But these ridges are marked by great summits and mountain masses ranging from 3,000 feet to 4,000 feet in height, such as Ben-y-Gloe and Ben Dearg among the Athol Mountains, and in the more extensive group of the Breadalbane Mountains, the peaks of Schiehallion, Ben Lawers, and Ben More. The lesser heights of Ben An and Ben Venue are well known for the famous gorge of the *Trossachs*, which is formed by their spurs at the approach to Loch Katrine from the Lowland Plain. Among these mountain ridges too are enclosed the greater number of the vast circle of Highland lakes, Loch Ericht, Loch Rannoch, Loch Tay, Loch Earn,

and Loch Katrine.   The whole of this district in fact,
with its massive mountains, its lakes, its torrents, and
its rocky glens, forms a scenery whose picturesqueness
is unrivalled in Britain.   But its wild and rugged
character leaves it barren and uninhabited, without a
single town of any note.   A few hamlets and villages,
indeed, lie here and there in the river valleys, and
are well known for the grandeur of the surrounding
scenery.   Such are *Blair Athol* and *Pitlochrie*, which
at either extremity of the Pass of Killiecrankie command the road to northern Scotland through the
mountains up the valley of the Garry.

The villages however increase in size as the
valleys widen to the Plain.   *Dunkeld* exactly marks
the spot where the Tay emerges from the mountains on the low sandstone flats : to the eastward
Alyth, Blairgowrie, and *Cupar Angus* in the basin
of the Isla, are all situated amid the fertile fields
of Strathmore ; while to westward the little town
of *Crieff* is placed in Strathearn.   The only really
large town, however, in the whole basin of the Tay
is *Perth*, the capital of the shire.   Lying on the
banks of the Tay, in the break between the Sidlaw
and the Ochill Hills, it commands the only means of
access to northern Scotland, whether by the plain
of Strathmore to the sea coast, or by the valleys of
the Tay and the Garry ; while by its position at the
head of the Firth it holds the key of the estuary.   The
great importance of its site made Perth in early times
the frequent residence of the Scotch kings, who were
crowned and held their court in the adjoining palace

of *Scone*. But in later days its military and political importance declined, and it is now but a moderate sized town with 25,000 inhabitants.

The remaining villages of Perthshire lie within the basin of the Forth, and are of very little note; they are *Dunblane* on the Allan, at the foot of the Ochill Hills, and *Doune* and *Callander* in the valley known as Menteith.

The whole population of Perthshire is comparatively small, amounting to 127,741 persons.

CHAPTER IX.

As we have already seen, the Lowland Plain consists of three parts, Strathmore, the basins of the Forth and Clyde, and Ayrshire. But Strathmore lies within the counties belonging to the Highland group, and Ayrshire within the counties of the Lowland Hills. We have therefore here only to do with that group of shires which falls within the basins of the Forth and the Clyde.

This group consists of ten shires, which altogether occupy an area of a little over 3,000 square miles, a space equal to about a sixth part of that occupied by the Highland group, and about 1,000 miles less in extent than the single shire of Inverness. The shires themselves are generally very small, only two of them being over 500 square miles in extent, while one covers but 36 square miles. But the geographical position of these midland counties gives them an importance which cannot be measured by mere size; for comprising as they do nearly the whole of the manu-

facturing and mining districts of Scotland, its chief
towns, its great sea-ports, nearly all its more important
shipping trade, and its best agriculture, they gather up
the great mass of the population and wealth of the
land.   With their 1,738,000 inhabitants they have a
density of population which is nearly ten times as
great as that of the Highland district, and a wealth
in proportion to their population and their industrial
resources.

The shires of the central plain differ greatly from
those of the Highlands in the mode of their formation.   Many of them are composed simply of districts
surrounding some town of note which had jurisdiction
over a part of the neighbouring land to which it gave
its name.   A glance at the map shows also the enormous
importance of the part played in their formation by
the two great rivers of the plain, for every large town
in this district lies upon the banks of one or the other
of these streams, and every county of the group communicates with the sea by the great estuaries which
they form.   The various counties thus naturally
group themselves according to the river basin in
which they lie, seven of them being included in the
basin of the Forth, and three in the basin of the
Clyde.

[A] COUNTIES OF THE FORTH.—As the Forth
flows out of Perthshire, the lower valley and estuary
of the river divide between the counties which lie
ranged along its northern and southern banks.   To
northward of the Firth lie *Fife*, *Kinross*, and *Clackmannan;* while to the south the river is bordered by

Stirling, Linlithgow, Edinburgh, and *Haddington,* these last being also known as the *East, West,* and *Middle Lothians.*

I. **Fifeshire** is composed of the peninsula which runs out to the North Sea between the Firths of the Forth and the Tay, being bounded on its land side by the Lomond and Ochill Hills, which part it from Perthshire, Kinross, and Clackmannan. The county has an area of 492 square miles, and is broken into two parts by the Lomond and the Fife Hills, a belt of igneous rocks which strike across it from west to east. In the northern part of the shire lies the fertile valley of the Eden, known also as the *Howe of Fife,* in the midst of which is situated the county town *Cupar;* while at the mouth of the valley on the coast is the sea-port town of *St. Andrew's,* the seat of an ancient archbishopric and the oldest University town in Scotland.

The southern part of Fifeshire, on the other hand, is made up of carboniferous rocks where coal is found; and to a manufacturing industry that has sprung up on these coal-measures, it adds a large shipping trade which it owes to its position on the Forth. The centre of the linen manufacture of the shire is at *Dunfermline,* a town with 15,000 inhabitants: to eastward of it lie *Kirkcaldy* and *Dysart,* two sea-ports on the Forth. The whole population of Fifeshire amounts to 160,735.

II. Westward of Fife is **Kinross,** a small tract of land only 73 square miles in extent, shut in between the Lomond and the Ochill Hills, and lying immediately around the little town of *Kinross* on the shore

of Loch Leven. It is hemmed in on every side by
Fife, Clackmannan, and Perth, and thus wholly shut
off from the Forth, while it merely touches the coal-
measures on its eastern border. Having thus no trade
nor especial industries its population is small, being
little over 7,000.

III. Yet further to the westward along the banks of
the Forth lies the little shire of **Clackmannan**, only
36 square miles in extent, formed by the district which
lies immediately round the town of *Clackmannan*, between the counties of Perth and Fife. From its position
within the limits of the coal-measures it has a large
manufacturing industry, and contains a population of
23,747 persons, chiefly employed in the coal and iron
mines, and in the making of woollen shawls and
tartans. *Alloa* on the Forth is the centre of its
mining industry.

IV. From Clackmannan we cross the Forth to enter
on the shires which border its southern bank, beginning with that of **Stirling**. This county stretches
along the whole upper course of the river from its
source to its estuary, and thus extends very nearly
across the peninsula of central Scotland. Its area is
equal to 447 square miles. In form it is somewhat
pear-shaped ; the thinner end runs up into the Highland district, stretching along the shores of Loch
Lomond between the shires of Dumbarton and
Perth, while the broader portion of the county lies
in the Lowland Plain, where it adjoins the shires of
Lanark and Linlithgow. This lower part of Stirling
is however very uneven in character, being broken by

a line of igneous rocks, the Campsie Fells, which
strike across its centre. To westward of them lies
a strip of Old Red Sandstone which in reality forms
a continuation of the valley of Strathmore; and to
eastward stretch the coal-measures of the central
plain.

In early times this easternmost portion of the shire
was from its geographical position one of the most
important districts in Scotland, forming as it did the
great pass into northern Scotland, and commanding
the head of the Forth, which was then the most important estuary of the country. Hence its chief town
Stirling, with its castle built on a steep rock rising
from the river-banks, held the passage of the Forth
as it flows between the Campsie Fells and the Ochill
Hills, and was practically the key of central Scotland,
and a military post of the first consequence. It was
thus that Stirling became in early times one of the
capitals of the kingdom. This ancient fortress however is now the centre of a great manufacturing district, and the town is largely employed in iron works
and woollen manufacture along with the neighbouring
towns of *St. Ninian's* and *Bannockburn* and many
lesser ones. *Falkirk* is an agricultural town on the
eastern border of the shire with large cattle markets;
lying at the head of the canal which unites the Firth
and the Clyde, it has a considerable trade with
Glasgow.

The population of the county, chiefly massed as
we have seen in its coal district, amounts to 98,218.

V. To the eastward of Stirling, and still bordering

the Firth of Forth, lie the three shires which occupy the district of the *Lothians*, known as the West, Middle and East Lothians.   In these shires we almost leave behind us the mining districts of the central coal-measures with their large industrial towns, and enter on a region of a very different character, where there is not a single manufacturing town, but where the only industry, that of tillage, is carried to its highest perfection.

The first of these shires, that of **Linlithgow** or *West Lothian*, is composed of a small tract of land of 120 square miles in extent which lies round the old town and royal palace of *Linlithgow*, so called from the little "lin" or lake on which it is built. Lying in a district which forms the best land for tillage in Scotland, while it is partially included within the limits of the coal-measures, the shire combines both agricultural and some little mining industries, and so maintains a large population of 41,000 inhabitants. Besides the chief town it contains the mining town of *Borrowstoneness* or *Bo'ness* on the Firth of Forth; and the agricultural town of *Bathgate*, lying on its eastern border.

VI.   The county of **Edinburgh** which borders Linlithgow on the east, and forms the district of *Mid-Lothian*, is in the same manner made up of the territory surrounding. the city of Edinburgh.   From the shores of the Forth it extends southwards in the form of a rough crescent bent round the extremity of the Pentland Hills: one horn stretches to the south-west between Linlithgow and Lanark; the other reaches to the south-east between Peebles and Berwick, and includes the Muirfoot Hills. From the Pentland Hills on the southern border of the shire outlying spurs are thrown out to the northward which reach as far as the Forth, where they terminate in the abrupt heights of Arthur's Seat and Castle Rock rising out of the valley of the little river Leith shortly before it pours itself into the Forth. On Castle Rock and in the valley below it lie the fortress and city of EDINBURGH, the capital of Scotland. The geographical position of this city was always one of great political importance, lying as it did where the main road from England, after skirting the Lowland Hills, opened into the centre of the Plain, and where the estuary of the Forth formed the highway of communication with Europe. Situated among rich agricultural plains, it is mainly important as a centre of government and the seat of a University, and is in a very slight degree an industrial city; and hence its population is less than half that of Glasgow, amounting to only 200,000 persons. Its seaport, *Leith* on the Forth, lies so close to Edinburgh as to form practically a suburb of the city, and contains 44,000 inhabitants. Other ports of less consequence line the shore of the estuary, such as *Musselburgh* and *Portobello*. The only inland town with as many as 5,000 inhabitants is *Dalkeith* on the eastern limits of the shire.

The whole area of the county is only 362 square miles, while it contains 328,379 inhabitants, a larger number for its size than any other shire in Scotland.

The bulk of its people, as we have seen, are gathered in the capital and by the borders of the Forth, where they find employment in the shipping trade. But the inland districts have also varied sources of wealth both in the rich agriculture which characterises the Lothians, and in the mines and manufactures of silk and wool due to the coal-measures that overlap the limits of the shire.

VII. The last county comprised within the basin of the Forth is **Haddington**, or *East Lothian*, which extends from the shire of Edinburgh eastward as far as the mouth of the estuary. Its southern boundary is formed by the Lammermuir Hills, which shut it off from Berwick, and form a steep wall of cliffs overlooking the plain that stretches from their base to the shore of the Forth, a plain broken by isolated hills such as North Berwick Law on the northern coast. Across the centre of this plain flows the river Tyne, having on its banks the little county town *Haddington*, which gives its name to the shire; and near its mouth the castle and village of *Dunbar*, situated on a rocky promontory jutting out into the North Sea.

The area of Haddingtonshire is 271 square miles; its rich agricultural plain, which is cultivated with great skill, supports a comparatively large population for a rural district, 37,770 persons.

[B.] THE COUNTIES OF THE CLYDE are three in number, but these three shires, Lanark, Dumbarton, and Renfrew occupy a tract of land only 400 square miles less in extent than that covered by all the

counties of the Forth basin. The Clyde does not, like
the Forth, form a dividing line which parts between the
shires on either side of it, for the greater part of its
valley lies right through the centre of Lanarkshire,
and it is mainly its estuary that separates between the
counties of Renfrew and Dumbarton. The basin of
the Clyde, with the counties which it includes, is also
distinguished from that of the Forth by containing
the more important part of the coal-measures of the
central plain, and represents therefore the manufacturing and commercial activity of Scotland to a far
greater degree than the basin of the Forth, whose
industry is mainly agricultural.

I. **Lanarkshire**, the largest county of the Lowland Plain, occupies the centre of this plain, lying
wholly inland midway between the eastern and
western seas. It coincides exactly with the basin of
the Clyde from its source to near the head of its
estuary. To the north and east the boundaries of
the shore lie along the line of waterparting that runs
between the rivers of the Atlantic and the North Sea:
for where it touches Dumbarton, Linlithgow, and Haddington, Lanarkshire is parted from the basin of the
Forth; and where it marches with Peebles from the
basin of the Tweed. On the south the waterparting of
the Lowland Hills divides it from Dumfries and the
watershed of the Solway Firth; while to the west a
ridge of igneous rocks cuts it off from the basins of
the Ayrshire rivers.

The shire thus clearly defined for us by the river-
basin in which it lies, has an area of 889 square miles,

an area equal to nearly half the extent of the counties of the Forth. It is composed of parts of very different character. The upper valley of the Clyde, or *Clydesdale*, has the form of a wedge driven into the midst of the Lowland Hills, and shares in their characteristic scenery—a waste uncultivated tract of country with smooth rounded hills and wide bleak pasture lands. From this pastoral district the river enters by a series of rapid Falls on its lower agricultural valley, where just below the Falls the little county town *Lanark* lies amid cultivated hollows of orchards and cornfields, dotted with thriving villages. These, however, give place as the river passes on to the busy life of a manufacturing district, where the valley is marked by coal-pits, chimneys, blast furnaces, a dense population, and great towns darkened under their perpetual pall of smoke.

On the left bank of the river lies *Hamilton*, at the junction of the Avon with the Clyde ; to the eastward of the stream are the towns of *Wishaw* and *Motherwell*, with *Airdrie* and *Coatbridge* on a little affluent of the Clyde. Close to the estuary, however, there lies a far greater town than these, GLASGOW, the centre of a host of lesser ones of which *Rutherglen*, *Govan*, and *Partick* lie within the limits of the shire, while many more gather along the borders of neighbouring counties where they form a group of dependant manufacturing towns. The centre of this group, Glasgow, is the third greatest city of the British Islands, with a population of 480,000 persons, and far exceeds any other town in Scotland in size and

wealth. Its factories, between 3,000 and 4,000 in
number and employing about 100,000 workmen, are
engaged in a remarkable variety of manufactures, for
Glasgow has its cottonspinning like Manchester, its
silk-weaving like Macclesfield, its clothmaking like
Leeds, its manufacture of jute like Dundee, its
potteries like Worcester, its metal-works like Birmingham, and its shipbuilding and fitting out of
steam vessels like Newcastle. To these and other
lesser industries it adds another advantage in which it
resembles Liverpool, that of being the natural centre
of trade and commerce for the whole surrounding
district, and in fact for the whole of the Lowland
Plain even as far as Falkirk eastward. Into Glasgow, as we have seen, all the wealth of the plain is
poured, and through its port nearly all the foreign
commerce of Scotland passes—a commerce which is
fed by the inexhaustible resources of the New World
towards which the estuary of the Clyde opens. With
this wealth of industrial resources, Lanarkshire contains 765,340 inhabitants, that is, in its area of less
than 900 square miles it has a population equal to
two-thirds of the population of the vast district which
constitutes the group of the Highland counties.

II. The Clyde, which traverses the centre of Lanarkshire, widens to its estuary soon after passing out
of the limits of that county, and its course then forms
the line of division between two lesser shires. Along
the northern banks of the estuary stretches the county
of **Dumbarton**, formerly known as the *Lennox*, a
county of only 241 square miles in extent, and of very

irregular form. The bulk of it consists of the district that immediately surrounds the town of *Dumbarton*, whose famous castle on a steep rock of 500 feet in height stands where the Leven unites its waters with the Clyde, and thence overlooks and guards the Firth. This commanding position made of Dumbarton the capital of the ancient kingdom of Strathclyde.

From the bank of the Firth a long thin wedge of country extends northwards into the Highland district between the waters of Loch Long and Loch Lomond, lakes which sever this part of the shire from the counties of Argyle and Stirling to west and east. The town of *Helensburgh*, situated on a sea loch opening from the Firth of Clyde, here marks the meeting point of the plain and the southernmost spurs of the Highland rocks. The shire of Dumbarton also possesses an outlying tract of land imbedded between Stirling and Lanark, and included within the mining district; this detached fragment is marked by the manufacturing town of *Kirkintilloch*.

The position of Dumbarton on the estuary of the Clyde, and its neighbourhood to the manufacturing centre of Glasgow, give it a comparatively large population of 58,857 persons.

III. The southern bank of the estuary of the Clyde, opposite to Dumbarton, is bordered by a tract of low-lying land which forms the shire of **Renfrew**. With an area of 247 square miles, it extends from the river side southward to the crest of a low ridge of igneous rocks which part it from Ayrshire, and stretches westward to the borders of Lanarkshire.

The county of Renfrew is the most thickly peopled district in all Scotland, containing within its narrow limits of 245 square miles, nearly 217,000 inhabitants. This it owes to the advantages of its geographical position. Its eastern portion, lying within the coal-measures, and bordering so closely as it does on Glasgow, the centre of industrial activity in Scotland, is the seat of extensive manufactures, such as those of *Pollockshaws*, *Barrhead*, *Johnstone*, and *Paisley*, the most important of all, with its factories for wool, silk, cotton, muslin, and shawls, which are famous in all parts of the world.

On the other hand, the western half of Renfrew, which lies along the estuary, includes the great shipbuilding docks for which the Clyde is noted, and some seaport towns which lie near the mouth of the river, such as *Port Glasgow*, *Greenock*, and *Gourock*.

The two largest towns of Renfrew—Greenock and Paisley—are among the greatest towns in Scotland, for their population of 57,000 and 47,000 give them a place among the first six towns of the country. In importance they rank far before the little county town *Renfrew*, with its 4,000 inhabitants, which lies midway between them.

CHAPTER X.

THE eight counties which occupy the slopes of the Lowland Hills, and thence overlap in parts the borders of the plain, cover an extent of country equal to over 5,000 square miles. In general character they are all alike, consisting mainly of the tracts of pastoral upland which constitute the hill country, broken by narrow river-valleys and glens where, sheltered by the bare smooth slopes on either side, a little tillage is carried on. So solitary and thinly-peopled is the whole district that throughout the length and breadth of the Lowland Hills there are but four towns with over 5,000 inhabitants. The contrast which they present to the Lowland Plain is strikingly shown in the shire of Ayrshire, which extends northwards into the low country, and there within a very narrow compass contains as many as three important towns, of which two are larger than any in the Lowland Hills.

The main waterparting of Scotland, as it strikes across the Lowland Hills from north to south, so divides between the shires which lie to right and left of it as to break them into two groups corresponding to the two watersheds. (*a*) The four south-western counties, those of Ayr, Wigton, Kirkcudbright, and Dumfries, drain their waters into the western sea. (*b*) The four north-eastern shires, on the other hand, Peebles, Selkirk, Roxburgh, and Berwick, drain their waters into the eastern sea. These two groups we shall therefore consider separately.

[A] THE SOUTH-WESTERN COUNTIES occupy a district twice as great in extent as those of the north-east, and are all maritime shires.

I. The first of these counties, **Ayrshire,** is a semi-circular tract of country bordering the Firth of Clyde, and shut in by hills on the land side where its boundaries run along the crest of the heights that enclose it. To the north-east a low belt of igneous rocks parts it from the counties of the Plain, Renfrew and Lanark; while to the south-east the mountains of Black Larg, Merrick, and the Carrick Hills, divide it from the Hill counties of Kirkcudbright, Dumfries, and Wigton. From these heights are thrown down a number of small streams, such as the Ayr, the Doon, and the Irvine, which traverse the shire and pour themselves into the western sea.

Ayrshire itself, with its area of 1,128 square miles, is the largest shire of southern Scotland. It is broken into three districts with clearly-defined differences of character. To the northward, *Cunninghame* stretches beyond the Lowland Plain with its coal-measures into the hilly tract of the igneous coast ranges; *Kyle*, in the centre of the shire, occupies the border land where the Plain and the Hills meet; while *Carrick*, in the south, stretches up into the Hill district, and is marked by all the features of its wild upland scenery.

The chief centres of population naturally lie in the low ground on the coal-measures, where the seaport of *Ayr*, the county town, with nearly 18,000 inhabitants, and the lesser port of *Irvine*, form the outlets of the manufacturing district that lies behind them, and of which the capital is *Kilmarnock*, with a population of 23,700. In fact, so thickly are the people massed together round the mines and factories of the coal district that Ayrshire with 200,000 inhabitants contains little less than half the whole population gathered in the shires of the Lowland Hills.

II. The three remaining counties of the south western group lie on the southern watershed of the Lowland Hills, and drain their waters into the Solway Firth.

Wigton, the smallest of these shires, with an area cf 485 square miles, is made up of two irregular peninsulas. One of these, the Rhinns of Galloway, projects into the North Channel between Loch Ryan and Luce Bay, and its westernmost point, marked by *Portpatrick*, is but twenty-two miles distant from the opposite coast of Ireland. The second peninsula, which includes a great part of the Galloway Hills, extends southward into the Solway Firth. The surface of the county, with its low hills and reaches of rough moorland, is throughout of a barren and mountainous character; and hence its population is comparatively small, amounting only to 38,830. All its towns lie along the coast on little harbours. *Wigton*, the county town is hardly more than a village ; *Stranraer* on Loch Ryan is a port of somewhat more consequence.

III. To eastward of Wigton, and parted from it by the little river Cree, lies the much greater county of **Kirkcudbright**. This shire, with an area of 889 square miles, lies in the form of a rough square on the southern slopes of the Lowland Hills. Its north-western boundary, which adjoins Ayrshire, runs along the crest of the heights that form the waterparting from Merrick to Black Larg; while a belt of high ground thrown down from Black Larg to the Solway Firth near Criffel, severs it from Dumfries on the east as well as from the valley of the Nith.

The shire so inclosed is traversed right through its centre by the river Dee and its affluent the Ken, while to right and left of the valleys formed by these rivers rise long reaches of hill and upland. To westward lie the undulating lines of the Hills of Galloway; and to eastward the ground rises to the crest of the uplands which mark the western limits of the valley of the Nith. The broad moorland solitudes which thus constitute the bulk of the shire form one of the most thinly-peopled districts in the Lowlands : even in the valley of the Dee there are but two villages that contain a little over 2,000 inhabitants, *Kirkcudbright*, the county town on the river estuary ; and

Castle Douglas. The population of the whole shire is less than 42,000.

IV. The county of **Dumfries**, which lies to the east of Kirkcudbright, occupies that great tract of the southern watershed, over 1,000 square miles in extent, which comprises the valleys of Nithsdale, Annandale, and Eskdale. Its northern and eastern limits are accurately marked by the high belt of ground that forms the central ridge and waterparting of the Lowland Hills, for the sinuous line which passes over the crest of the Lowther Hills and Queensberry Hill to the north of the shire and severs it from Lanarkshire, severs it too from the basin of the Clyde; while to the north-east and east, where Hart Fell, Ettrick Pen, and Wisp Hill mark the boundaries of Peebles, Selkirk, and Roxburgh, they also mark the limits of the basin of the Tweed.

Shut off in this manner alike from the river basins of the north and east, Dumfries is itself traversed, as we have seen, by all the principal streams which empty their waters into the inner part of the Solway Firth. The western half of the county forms the basin of the Nith, at whose mouth lies the county town, *Dumfries*, which with its 15,450 inhabitants is one of the most populous places of the Lowland Hills. Though situated nine miles above the mouth of the river, it yet forms a seaport of some consequence owing to the daily tides which carry vessels into its harbour.

The valley of the Annan, in the eastern half of the shire, corresponds to that of the Nith in the western, and is parted from it by the Lowther and Queensberry hills.   The picturesque little vale of *Moffatdale*, which opens into it under Hart Fell, is noted both for its wild scenery with deep dark glens and defiles, and for its mineral springs. At the mouth of the Annan lies a small port, *Annan*, not a fifth the size of Dumfries.

The Esk as it takes up the waters of the Liddel marks the point where Dumfries touches English soil.   Near its mouth lies the low reach of marsh known as *Solway Moss*, which by laborious cultivation and drainage has now been rendered fertile soil.

The population of Dumfries, over 74,800, is comparatively large for a shire in the Lowland hills.

[B] THE NORTH-EASTERN COUNTIES form also a group of four shires, which however are only equal in extent to about one-half of the south-western group, and differ from those counties in the fact that they lie for the most part inland.

These four shires are closely linked together by one great river, the Tweed, whose entire course lies among them, and into whose channel almost all their waters are poured.   From its source on the western border of *Peebles*, the Tweed passes eastward across *Selkirk*, and thence winds along the borders of *Roxburgh* and *Berwick* to its estuary, taking up through Roxburgh the drainage of the Cheviot Mountains, and through Berwick that of the Lammermuir Hills.   Thus the grouping of these shires lying on the borders of England closely resembles the arrangement of the English county groups.

I. **Peebles**, part of which coincides with the older district of Tweeddale, forms the upper basin of the

Tweed. By Hart Fell to the south it is severed from the watershed of the Solway Firth and from Dumfries : by the Pentland hills to the north it is cut off from the basin of the Forth and Edinburghshire ; to westward Culter Fell and the main waterparting line divide it from Lanarkshire and the basin of the Clyde ; and to eastward the heights stretching from Hart Fell by Broad Law to Dollar Law mark the border of Selkirk, into which county the river finally breaks as it passes out of Peebles.

The upper valley of the Tweed, thus closely hemmed in between the heights which bound it to east and west, is of a wild and barren character, a solitude of moorland and narrow mountain glens fitted only for sheep-pasture. Within its area of 354 square miles it contains but a scanty population of little over 12,000 people, and its county town, *Peebles*, lying on the Tweed as it bends eastwards between Dollar Law and the Muirfoot hills, is a mere village.

II. As the Tweed passes out of Peebles it enters on the shire of **Selkirk,** whose northern portion it traverses on its way eastward to Roxburgh. Selkirk, which coincides with the wild district of Ettrick Forest, consists mainly of the basin of the Ettrick, an affluent of the Tweed thrown down from Ettrick Pen on the southern borders of the shire where it adjoins Dumfries. Parallel to Ettrickdale runs a second glen, that of Yarrowdale, in which lie St. Mary's Loch and the stream of the Yarrow, which empties itself into the Ettrick close to the little county town, *Selkirk*. A narrow belt of land that lies along the northern banks of the Tweed and adjoins Edinburghshire contains the manufacturing town of *Galashiels*, where woollen goods and tweeds are largely made, and whose population of nearly 10,000 makes it one of the most considerable towns of the Lowland Hills.

The whole population of the shire is about 14,000. If from this number we take away the 12,000 comprised in the two towns above mentioned, we shall realise the solitude of the wide moorland reaches, 257 square miles in extent, which form the county of Selkirk.

III. The Tweed as it flows eastward from Selkirk divides between the counties of **Roxburgh** and Berwick. From the river valley Roxburgh extends southward to the English border, with which it marches for the whole length of the Cheviot Mountains, from the Liddel to near the mouth of the Tweed.

The shire is broken into two unequal parts by a belt of uplands marked by Wisp Hill, Peel Fell, and Carter Fell, which connect the Lowland hills with the Cheviot mountains, and form the southern extremity of the Scotch waterparting. On the western watershed of these hills lies Liddisdale, through which the Liddel passes into Dumfries and so to the Solway Firth. The eastern watershed is much more extensive, and comprises the greater valley of Teviotdale, traversed by the Teviot as it carries the drainage of the surrounding hill country to the Tweed.

Teviotdale naturally forms the centre of population of the shire, since it adds to the pastoral industry of the hill-slopes on either hand a considerable manufacture of wool and hosiery. Its chief manufacturing town, *Hawick*, has 11,000 inhabitants; the county town,

Jedburgh, however, lower down the river, is little better than a village, and is even smaller than *Kelso* with its 4,000 inhabitants, which lies at the junction of the Teviot with the Tweed. *Melrose*, to the westward of it, is a village with some famous abbey ruins, situated where the Tweed winds round the solitary group of the Eildon Hills.

The entire population of Roxburgh, in an area of 665 square miles, amounts to nearly 54,000 persons.

IV. On the northern banks of the Tweed, opposite to Roxburgh, lies the shire of **Berwick**, the only maritime county of this group. To westward it touches the shire of Edinburgh, and to northward is parted from that of Haddington by the Lammermuir hills, whose southern slopes extend into Berwick, where they are furrowed by the streams that constitute the chief northern affluents of the Tweed. Of these the more important are the Whiteadder and the Lauder, the latter of which gives to western Berwick its name of Lauderdale. The county is famous for its great fertility and excellent farming, but it contains no large town. Berwick-on-Tweed, lying on the English border, was once a port of great consequence, and formed the ancient capital of the shire, but having been taken from Scotland by the English, it was constituted into a separate district independent of the Scotch kingdom. The present county town of the shire, *Greenlaw*, is a mere village, with 800 inhabitants; not far from it lies a second larger village, *Dunse*.

Nearly 36,500 people are comprised within the limits of Berwick, with its area of 460 square miles.

County.	Chief Towns.	Area in Sq. Miles.	Population.
Counties of the Highlands.			
Orkney & Shetland	Kirkwall . .	935	62,877
Caithness . . .	Wick . . .	712	39,989
Sutherland . .	Dornoch . .	1,886	23,686
Ross & Cromarty	Dingwall . .	3,151	80,909
Inverness . . .	Inverness .	4,255	87,480
Argyle	Inverary . .	3,255	75,679
Bute	Rothesay. .	217	16,977
Nairn	Nairn . . .	215	10,213
Elgin	Elgin . . .	531	43,598
Banff	Banff . . .	686	62,010
Aberdeen . . .	Aberdeen. .	1,970	244,607
Kincardine . .	Stonehaven .	394	34,651
Forfar	Forfar . .	889	237,528
Perth	Perth . . .	2,834	127,741
		21,930	1,148,145
Counties of the Lowland Plain.			
Fife	Cupar . . .	513	160,310
Kinross	Kinross . .	78	7,208
Clackmannan . .	Clackmannan	36	23,742
Stirling	Stirling . .	462	98,179
Linlithgow . . .	Linlithgow .	127	41,191
Edinburgh . . .	Edinburgh .	367	328,335
Haddington . .	Haddington .	280	37,770
Lanark	Lanark . .	889	765,279
Dumbarton . .	Dumbarton .	320	58,837
Renfrew . .	Renfrew . .	247	216,919
		3,319	1,737,770

Country.	Chief Towns.	Area in Sq. Miles.	Population.
Counties of the Lowland Hills.			
Ayrshire . . .	Ayr . . .	1,128	200,809
Wigton	Wigton . .	485	38,830
Kirkcudbright .	Kircudbright	898	41,859
Dumfries . . .	Dumfries. .	1,063	74,808
Peebles	Peebles . .	354	12,300
Selkirk	Selkirk . .	257	14,005
Roxburgh . . .	Jedburgh . .	665	53,974
Berwick . . .	Greenlaw .	460	36,486
		5,310	473,071

IRELAND.

CHAPTER I.

IRELAND, the second of the two islands which make up the British group, lies off the western shores of Great Britain, and is bounded to north, south, and west by the Atlantic Ocean, which stretches away from it to the westward in an unbroken expanse of sea for nine thousand miles to the continent of America. It thus forms the extremity of Europe, and the last fragment of that great tract of land which stretches eastward across Asia to Behring's Straits. From its neighbour island, however, it is parted only by narrow straits. The *Irish Sea*, which divides it from England, opens out to north and south by two channels into the Atlantic Ocean : these are the *North Channel*, which runs between Ireland and Scotland, and is at its narrowest part only thirteen miles wide ; and St. George's Channel, which flows between Ireland

IRELAND

By Keith Johnston F.R.S.E

English Miles

0   10   20   30   40

U

D

Tory I.

Bloody Foreland

N. Aran I.

Gweedore B.

Loughros M.

Rossea I.

SLIGO

Donegal BAY

Sligo B.

Killala B.

Downpatrick Hd.

C

Broad Haven

Erris Hd.

N A

W & A K Johnston Edinburgh and London

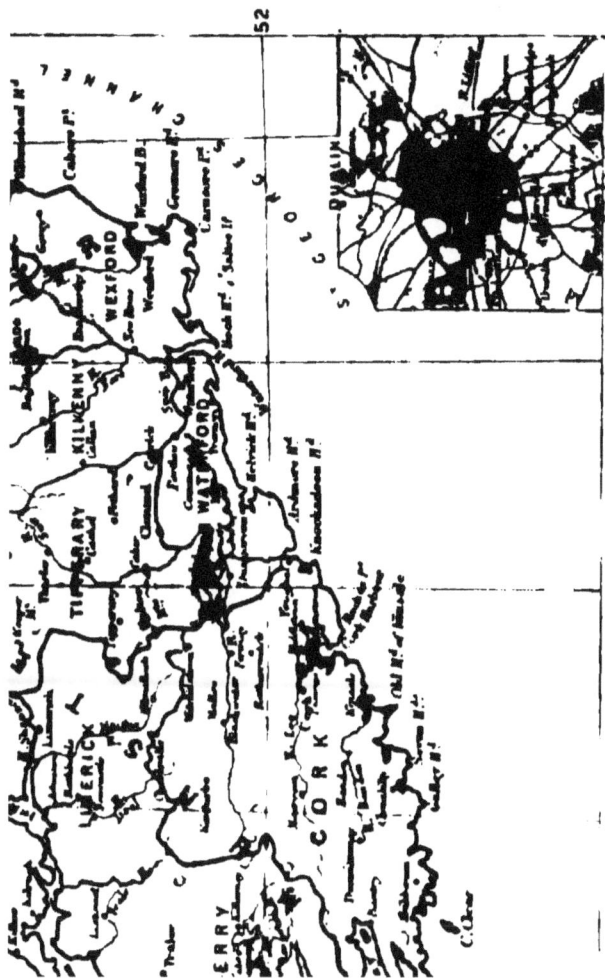

Longitude West 9 of Greenwich 8 7 6

52

ST GEORGES CHANNEL

WEXFORD
KILKENNY
WATERFORD
TIPPERARY
LIMERICK
CORK
KERRY

and Wales, and is four times as broad as the first. Of the general characteristics of Ireland, in size and physical structure, we have already spoken (see pp. 8, 9), and we have noted how great an influence is exerted by its position on the character of its population, as well as on its climate (see p. 5), and through its climate on the nature of its agriculture. Here, therefore, we may pass at once to survey in detail its inner characteristics, the nature of its mountains and plains, the extent of its river valleys, and the general type of its coast.

MOUNTAIN MASSES.—In form Ireland is broader, shorter, and more compact than its fellow island of Britain. Roughly speaking, it is something like a lozenge set corner-wise in the ocean, having its four sides marked by lines drawn between the headlands of Fair Head in the north-east, Erris Head in the north-west, Mizen Head in the south-west, and Cornsare Point in the south-east. Within the space which these lines inclose lies a country wholly different in structure and internal character from England or Scotland. It is in effect a great plain, a plain broken here and there by low hills, and only rising towards the sea-coast into mountain ranges of a date geologically far more ancient than the plain itself. We can best realise the character of this mountain ring by supposing the land to sink down a little into the sea, as has happened already in past ages. In such a case Ireland would form a vast lake dotted with islands, surrounded by a ring of land formed by the mountains which now border the coast, and with but two openings to the sea, where breaks occur in this mountain-ring, one at Galway Bay, and a much wider one at Dublin Bay. The different parts, however, of the encircling hill-ranges vary much in character. On the north and on the south they widen into a broad belt of highlands, along the west they run in a narrower border-line, while on the eastern side of the county two clusters of mountains stand like massive portals on either side of the level reach of shore about Dublin Bay, an opening which forms a huge gateway to the central plain from the side of Europe.

The Northern Highlands of Ireland are made up of the mountains of Antrim on the north-east and of Donegal on the north-west. The *Mountains of Antrim* constitute a broad high tableland, ending abruptly on the sea-coast in precipitous cliffs, and broken by steep hills, such as *Mount Trostan* (1,800 feet) near the sea. The *Mountains of Donegal* form a much higher and bolder mass of mountains cut by steep valleys running from north-east to south-west. Many of them have an elevation of 2,000 feet ; their greatest height is reached in *Mount Errigal* (2,462 feet), while to the south-west *Slieve League* rises sharply almost out of the sea to nearly as great a height as Errigal.

The Antrim and Donegal Mountains are connected by the lower ranges of the *Carntogher* and *Sperrin Mountains*, from which outlying lines of hills run southwards, so that the whole of northern Ireland above a line drawn from Dundalk Bay to Sligo Bay has a more or less mountainous character.

The **Western Highlands.**—The northernmost of
these, the mountains of *Connaught*, form two peninsulas
projecting into the Atlantic between Sligo and Galway
Bays, and parted from each other by Clew Bay.
The northern peninsula consists of the *Mountains of
Mayo*, the chief of which are the *Nephin Beg Mountains*, yet higher and wilder than those of Donegal;
these run westward from their chief summit, *Nephin*,
2,646 feet high, and terminate in the bold cliffs of
Achill Island, which reach a height of from 900 to
1,800 feet. A range of lesser elevation runs north-
eastward to Downpatrick Head. The southern peninsula consists of the *Connemara Mountains*, which reach
their greatest height near the coast, and lie sometimes
in isolated masses of granite, like *Croagh Patrick*,
and *Mulrea*, and sometimes cluster in groups like
the *Twelve Pins of Bunnabeola*.

Passing southwards by the scattered heights of
Clare, we reach a south-western group formed by
the *Mountains of Kerry*. These consist of three
peninsulas, parted from each other by Tralee, Dingle,
Kenmare, and Bantry Bays. In the northernmost of
these peninsulas, *Brandon Hill*, the second in height
among Irish mountains, rises to over 3,000 feet; the
central peninsula is occupied by a range in which we
find the highest peaks in Ireland, the *Macgillicuddy
Reeks*, whose loftiest summits reach 3,400 feet; while
the southernmost is composed of the mountains of
Glengariff.

The **Southern Highlands.**—In the south, as in
the north, the mountain-ring widens into a broad belt

of hilly country, made up for the most part of parallel ranges running eastward and westward between the coast and the central plain. The Kerry Mountains are prolonged eastward by the *Bochragh Hills*, whose line is continued across the valley of the Blackwater by the *Knockmealdown Range* to the hills of *Waterford*. North of this line runs the parallel range of the *Galtees*, and again still further to the north the *Silvermine* and *Slievebloom Hills* stretch away north-eastward along the bounds of the central plain.

The Eastern Highlands.—These are composed of those two circular masses of granite mountains which we have already described as the portals to the great gateway of the west. The southern group, that of the *Wicklow Mountains*, is the larger and more remarkable of the two. Its greatest height, *Lugnaquilla* (over 3,000 feet), is loftier than any of the Irish mountains except the two chief summits of Kerry, and like the Kerry mountains, this group contains lakes and waterfalls which have made it only second to the mountains of the south-west in the beauty of its scenery.

A long range of low shore which stretches to the north of Wicklow parts its mountains from a second mass of granite rocks, the *Mourne Mountains*. These form a small compact group projecting into the sea between the bays of Dundalk and Dundrum. Their highest summit, *Slieve-Donard*, rises 2,796 feet above the sea, and is surrounded by a number of peaks which exceed 2,000 feet in elevation.

THE PLAIN.—The great plain of limestone rock

which is girded about with these mountains was
once covered with dense forest and is still luxuri-
antly wooded.   It is now an immense pasture-field,
which gives to Ireland its name of the Emerald Isle.
Its surface is scarcely broken by a few low hills, and
the only interruption to its abundant fertility are the
tracts of bog land which lie across its centre.   Two
lines drawn across the island from Dublin to Galway,
and from Wicklow to Sligo, will take in between them
the chief part of this bog-land, which forms not one
morass, but a number of small bogs separated by
ridges of cultivated land, and provides the turf used
for firing in the absence of coal.   The plain itself is
traversed by but one large river, which gathers up
into its basin a multitude of little streams, by which
the lowland is watered; but this river, the *Shannon*,
is the greatest in the British Islands, and from its
mouth almost to its source forms a highway for trade
into the heart of the land.

LAKES.—Besides this vast plain, Ireland possesses
another characteristic feature in the multitude of its
lakes.   The bulk of these lie in the plain itself, and
among the lower slopes of the hills, and extend over
one half of the island, that half which lies to the
north-west of a slanting line drawn from Belfast
Lough to Kenmare Bay.

In this feature Ireland resembles that part of Britain
which is most opposed to it in structure, the Highlands
of Scotland; but in their distribution and relation to
each other its lakes differ widely from those of either
Scotland or England, for while the latter lie isolated

from each other in lines radiating from some central point, the Irish lakes form long chains of water linked to each other by connecting rivers.

They may be divided into five distinct groups.

(1.) **Lough Neagh** lies all by itself below the Antrim Mountains; extending as it does over an area of 150 square miles it is the greatest inland lake in the British Islands, while in the whole of Europe there are but three larger than it.

(2.) The **Lakes of the River Erne** form a long chain running from the central plain to Donegal Bay. The chief of these are the *Upper* and *Lower Erne*, connected by the river of the same name.

(3.) The **Lakes of the Shannon** are a long series of waters beginning near Sligo Bay, and linked together by the Shannon, to whose estuary they extend. The greatest among these are *Lough Allen*, *Lough Rea*, and *Lough Derg*.

(4.) The chief **Lakes of Connemara** are *Lough Mask* and *Lough Corrib*, which are joined together by an underground channel.

(5.) The **Killarney Lakes** are three in number, the *Upper*, *Middle*, and *Lower* lakes, all of which open into each other. They are extremely beautiful in scenery, lying as they do at the feet of the highest mountains of the island, the Macgillicuddy Reeks.

THE RIVER SYSTEM of Ireland is wholly unlike that of Britain in its character. In Britain the rivers rise in the western heights, and crossing the centre of the country fall into the eastern sea. But in Ireland, owing to the encircling girdle of mountains, the rivers have a very different course. Having their sources in these bordering heights, they fall into the sea on the same side of the island on which they rise, without ever crossing the central plain. This plain is in fact only traversed by the one great exception to the general rule, the Shannon, and by the small streams which form its tributaries.

The Rivers of the West.—The *Shannon*, the greatest river of the British Islands, rises in the lowest outskirts of the broad belt of hills which crosses northern Ireland, passes through the central plain in a long series of lakes, taking up only one important tributary, the *Suck*, from the western heights, and at last, after crossing the most fertile soil of the country in a course of 224 miles, opens into the Atlantic by a magnificent estuary sixty miles long, and from one to eleven miles broad. It is the only navigable river of Ireland, since it alone has a course long enough to allow it to grow to any size; and its waters are traversed by a number of steam vessels carrying passengers and the cattle of which its chief trade consists. On the same western coast with the Shannon are two small rivers having their course in an opposite direction to the north, the *Moy*, which falls into Killala Bay, and the *Erne*, which passes into Donegal Bay.

The Rivers of the North are few and small: the *Foyle*, which enters into Lough Foyle; and the *Bann*, which rises in the Mourne Mountains, passes through *Lough Neagh* and falls into the Atlantic.

The Rivers of the East are equally unimportant:

they are the *Lagan*, which has a very short course to
Belfast Lough : the *Boyne*, whose course lies midway
between the Mourne and Wicklow Mountains, and
which falls into the sea at Drogheda, where it forms
a small harbour : and the *Liffey*, at whose mouth
stands the capital of Ireland, Dublin.

The Rivers of the South are more important,
for each one of these forms at its mouth a valuable
harbour. The easternmost of these are the *Slaney* and
the *Barrow*, which flow in parallel courses due south,
and open out into the estuaries of Wexford Haven and
Waterford Harbour. The valley of the Barrow forms
the dividing line between the Wicklow Mountains of
the east coast and the ranges of the southern coast ;
from the westward it takes up two tributaries, the *Nore*
and the *Suir*. The course of the Suir is remarkable,
since the lower valley of the river lies due west and
east, and thus exactly follows the direction into which
all the remaining southern rivers are thrown by the
parallel mountain ranges of this coast.

The *Blackwater*, the *Lee*, and the *Bandon*, all rise
among the mountains of the south-west, and flow due
east to the sea through the valleys which run between
the mountain ridges. They form at their mouths
Youghal Harbour, the magnificent harbour of *Cork*,
and *Kinsale* Harbour.

COAST-LINE.—The same general features may be
traced in the character of the Irish coast-line as in
that of England and Scotland, but there are some
important differences. (1.) The northern and western
shores of Ireland are fringed by a border of islets,

but these being for the most part mere barren rocks, are too small to afford the shelter given to England and Scotland by their more important islands. (2.) There is the same contrast as exists in Great Britain between the eastern and western coasts, the western being fenced by a wall of precipitous cliffs, torn into rude and wild forms by the waves of a tempest-driven ocean; while the eastern is for the most part composed of low, monotonous, sandy shores, stretching along a sea at once shallower and tamer than the Atlantic. This eastern coast, however, differs from that of Great Britain in being twice broken by masses of high granite mountains. (3.) The western coast is cleft by deep bays and fiords capable of giving shelter to the largest ships; the eastern shore, on the other hand, is marked by shallow curves and inlets of the sea, and these, not lying open like the bays of the west to the dash of the Atlantic waves, are half choked by shoals, sandbanks, and sunken rocks. (4.) A greater number of rivers empty themselves into the eastern sea than into the western; but these rivers, unlike those of Britain, are too small to form good harbours at their mouths, and the river of the west, the Shannon, is alone of importance.

On the whole, therefore, Ireland like Great Britain has its coasts deeply indented by the sea, so that the entire length of its shore is over 2,000 miles, and no part of the country can be more than fifty miles from good navigation. It has also for its size a greater number of natural harbours for large vessels than England or Scotland; but it is burdened by one disadvantage not

shared in by Great Britain, since its greatest ports are
turned away from the nearest land for trading purposes, and look out towards a vast ocean, which
during many hundred years of Irish history parted
Europe from the undiscovered lands of the New
World. The various ports which now exist on the
side nearest to England have been chiefly made by
artificial means, as the necessities of trade with the
only neighbouring country became more pressing.

The North Coast.—The northern coast is
much shorter than all the rest, and wild and rugged
in character like that of the west. It is cleft
by two deep inlets of the sea, *Lough Swilly*
and *Lough Foyle*, the latter of which forms a good
harbour. These are parted by a narrow mountainous peninsula ending in *Malin Head*. *Horn Head*
lies near the western extremity of the coast, while
Fair Head marks the eastern, only thirteen miles from
the Mull of Cantyre. Here the whole coast is formed
of volcanic rocks and beds of ancient lava : Fair
Head itself is formed of huge columns of old volcanic
rock, descending from the top of the cliff in steep
lines : *Rathlin Island*, hard by, is a continuation of
the same rocks : and the famous *Giant's Causeway*, a
little to the west, is a vast pile of perfectly regular
volcanic columns, projecting out into the sea for more
than 1,000 feet.

The West Coast.—The western coast of Ireland is marked by very distinct natural features. As
in England, three great mountain-masses project into
the sea, whose rock-walls, like those of Scotland, are

cut by deep bays and inlets which open to the south-west and run deep into the land.

Between the northernmost of these peninsulas, that of Donegal, and the central peninsula of Connaught, stretches a great inward curve of sea, which is broken along the coast into three smaller bays: *the Bay of Donegal*, shut in by mountains; the lesser but much more important *Bay of Sligo*, which opens on a fertile plain; and *Killala Bay*, lying to the west of it.

The peninsula of Connaught is divided into two parts by *Clew Bay*, with *Clare Island* lying at its mouth. The coast to the north of this bay is very irregular; its chief feature is the strangely-formed peninsula of *Mullet*, ending in *Erris Head*, which is attached to the coast by a strip of land but half-a-mile wide, and forms with the mountainous rocks of *Achill Island*, *Blacksod Bay*.

Between the peninsula of Connaught and that of Kerry lies a second great inward curve of sea, which is broken like the former into three smaller bays along the coast. The northernmost of these, *Galway Bay*, with the small *Arran Islands* lying across its opening, is an important harbour.

The *estuary of the Shannon* lies farther down the coast, between the promontories of *Loop Head* and *Kerry Head*. To the south of Kerry Head opens the small bay of *Tralee*. The peninsula formed by the Kerry Mountains in the extreme south-west is cleft by a number of deep bays and arms of the sea, nearly all of which afford excellent shelter for ships. The chief of these to the south of *Dunmore Head* are three : *Dingle Bay, Kenmare Bay,* and *Bantry Bay.* Small rocky islets are dotted along this rugged coast; of these *Valencia Island,* in Dingle Bay, is the only one which can be cultivated.

Of these many harbours, the estuary of the Shannon, with its port Limerick, is by far the most important. The greater number of the other bays are shut out by mountain ranges from easy communication with the central plain, and their trade is very limited, being often confined to fisheries alone.

The East Coast.—The eastern coast differs greatly from the western in its long regular line of low shore, broken only by two slightly projecting mountain masses, the *Mourne Mountains* and the *Wicklow Mountains.* Its central point is *Dublin Bay,* which is shut in to the north by the peninsula of *Howth Head,* and has at its south-eastern opening the artificial harbour of *Kingstown.* This bay, in spite of its sandbanks, has become of great import-ance for commerce, for Dublin is the great feeder of the country, and gathers in for the whole of Ireland the produce of foreign lands—tea, coffee, sugar, wood, wine, &c.

The bays lying to the north of Dublin are: *Dundalk* and *Dundrum Bays,* of little value for commerce; between them the smaller but more im-portant *Carlingford Bay;* to the north of these *Strangford Bay,* landlocked by the *Ards* peninsula, and forming the only good natural harbour of the eastern coast; and beyond it *Belfast Lough.* Belfast is, after Dublin, the great harbour of the east; but its trade lies, not in gathering goods into the country, but in sending out the home manufactures of linen and cotton. It is thus contrasted also with Limerick, whose export trade lies in live cattle. The lesser ports of the east export chiefly butter and farm produce.

To the south of Dublin there is but one bay, that of *Wexford Haven :* between it and Dublin the coast-line is only broken by a few rocky promontories— *Cahore Point, Wicklow Head,* and *Bray Head.*

The South Coast.—The southern coast from *Carnsore Point* to *Mizen Head* somewhat resembles the eastern in general character, but its bays, lying open to the Atlantic, are not choked by sand-banks, and its two chief harbours of *Waterford* and *Cork* are naturally far superior to the harbours of the east. Waterford Harbour, at the mouth of the Barrow, is sheltered by *Hook Head ;* its trade lies wholly in carrying cattle to Bristol. Cork, which is completely land-locked, gives safe shelter to the largest ships; it forms the only naval station in Ire-land, and through it passes the whole commerce with America, as well as a great cattle trade with England. The *Old Head of Kinsale* is a promontory guarding the small harbour of *Kinsale ;* and between *Cape Clear* and *Mizen Head* lie two or three natural ports capable of receiving large vessels, but only used for fishing-boats. *Cape Clear,* the southern point of *Clear Island,* is a mere mass of barren cliffs.

CHAPTER II.

THE physical differences between Ireland and Britain are reflected in the different character of their political geography. In Britain the counties group themselves in the river-basins of the country, and their boundaries are in many cases determined by its lines of waterparting. The commerce, too, and industry which come of the mineral wealth of Britain have made its towns the most important element of its social and political life. But in Ireland we find none of the physical conditions which produce these results. Its river-system, save for the Shannon, plays little part in its geography; mineral wealth it has none, and its commerce and industry, save at one point in the north, remain insignificant. Its shires, therefore, and its groups of shires, do not correspond with the areas of river-basins: nor have they been formed by the force of physical circumstances. They represent for the most part either the territories of old Irish tribes, or artificial divisions made by their English

conquerors; and the new *Provinces* into which they
are grouped correspond roughly with the areas of
kingdoms into which the island was split up while
it remained independent. Before the invasion of the
English Ireland was divided into five kingdoms—
Ulster, Connaught, Munster, Leinster, and Meath, the
two last of which subsequently became one. These
kingdoms still survive in the four Provinces into which
Ireland is broken up. The modern Ulster lies north of
a line from Donegal Bay to Carlingford Bay; Connaught stretches along the western coast from Donegal
Bay to Galway Bay, and reaches inland as far as
Leitrim and the line of the Shannon; Munster
includes the south-western and southern shore from
Galway Bay to Waterford Harbour, and its inland
border runs along the eastern limits of Tipperary to
Waterford; the remainder of the island along the
eastern coast from Waterford Harbour to. Carlingford
Bay forms the province of Leinster.

The relation of town and country is a yet
more remarkable distinction between the two islands.
Though the population of Ireland is twice as
great as that of Scotland, yet there are more large
towns in the Scottish Lowlands alone than in the
whole of Ireland. The great plain which forms
the bulk of Ireland is more thickly peopled than
almost any part of Europe of the same area; it is
dotted all over with quiet country towns and hamlets.
But all of these are small and poor, and there is not
a single inland town of 20,000 inhabitants. The really
large towns of Ireland are in fact but seven in

number—Dublin in the east, Belfast and Londonderr in the north, Limerick in the west, and Cork an Waterford in the south. But all of these lie on th sea-coast, and they form a thin fringe of sea-port which exist chiefly through their trade with England.

The four Irish Provinces are again divided int counties, of which there are thirty-two in all.

I. ULSTER has an area of 8,555 square miles, an a population of 1,830,398; in proportion to its siz it is therefore the most densely peopled of the Irisl provinces. By population and industry it is divide into two distinct parts. The half lying to the west i peopled by the ancient Celtic races who fled to th farthest mountains for shelter from the English in vasion, and have here preserved their old language In this part the industry is wholly pastoral anc agricultural. The half lying to the east has beer colonised by the Scotch, and is the only manufacturing district in Ireland; to its trade in linen and cottor Ulster owes the upgrowth within its limits of two o the great towns of Ireland, Belfast and Londonderry It contains nine counties, four of which lie along the coast.

(1.) **Donegal,** the north-western part of Ireland consists for the most part of a rugged mass o mountains rising steeply from the Atlantic Ocean The county is bordered by the sea from the Bay o Donegal to Lough Foyle. On its land side it is partec from Fermanagh by Lough Erne and the river Erne and from Tyrone and Londonderry by the river Foylc and its affluents. The steep mountain ridges, partec

by deep and narrow valleys, which traverse it from
north-east to south-west, rise to their greatest height
near the western shore in Mount Errigal, while in the
north they run up between Lough Swilly and Lough
Foyle to Malin Head, the most northerly point of
Ireland. The country is barren and scarcely cultivated,
the river-valleys being dotted here and there with
villages, such as *Lifford*, the county town, on the Foyle,
Letterkenny on Lough Swilly, *Ballyshannon* on the
Erne as it enters Donegal Bay, and *Donegal* on the
bay itself.

(2.) **Londonderry** is a triangle wedged in between Antrim and Donegal on the east and west, with
Tyrone to the south, and having an opening northwards
to the Atlantic. Lough Foyle and Lough Neagh lie at
opposite extremities of the county, whose boundaries
are marked to east and west by the rivers connected
with these lakes, the Foyle, which empties itself into
Lough Foyle, and the Bann, which comes from Lough
Neagh. The county is much broken by mountains and
hilly ground ; the chief heights, the *Sperrin Mountains*,
lie on its southern border, and the lesser *Carntogher
Mountains* cross its centre. Agriculture is therefore
very poor and scanty, but the linen manufacture
of the coast district has gathered to the shire twice
as great a population as is to be found in Donegal.
Londonderry, on the river *Foyle*, not far from the
Lough, is one of the two chief towns of northern Ireland, and possesses 25,000 inhabitants. *Coleraine* on
the Bann, and *Newtown Limavady* on the Roe, are
small towns near the shores of the same bay.

(3.) **Antrim** is a long narrow county on the north-western coast, and is shut in on every side by water, being bounded to north and east by the Channel that parts Ireland from Scotland, on its western side by Lough Neagh and the Bann, which sever it from Londonderry, and on the south by the Lagan and Belfast Lough, which part it from Down. Its mountains are chiefly built up of old volcanic rock, such as those which form the marvellous columns of the *Giant's Causeway*. The highlands of the county constitute a high tableland that extends along its eastern coast, and thence falls gradually to Lough Neagh and the Bann valley, where the land becomes fertile. But the population and in-dustry of the county mainly gather round the shores of the Belfast Lough, where the linen and cotton manu-factures of Ireland centre. At the head of the Lough is *Belfast*, the second greatest city in Ireland, with 175,000 inhabitants; on the north side of the same. Lough is *Carrickfergus;* and *Lagan*, on the river of the same name, lies at no great distance from its shores. The only town of any size that lies apart from this group is *Ballymena*, in the centre of the county.

(4.) **Down** is the fourth and last county of Ulster which rests on the sea. Its long coast-line is deeply indented by the Belfast Lough, Strangford Harbour, Dundrum Bay, and Carlingford Bay. On the land side it only touches two counties, Antrim and Armagh. Its southern portion is occupied by a mass of rounded granite heights, the *Mourne Mountains*, and the rest of the county is broken by scattered hills. The only cultivation consists in small farms or fields of flax, so that the linen manufacture is the main industry, and this gives employment to many small towns. The more important of these lie on the chief harbours, such as *Downpatrick* and *Newtown Ards* on Lough Strangford, and *Newry* at the head of Carlingford Bay.

(5.) **Tyrone**, with Armagh, Fermanagh, Monaghan, and Cavan, form the five inland shires of Ulster. Tyrone stretches right across the centre of this province from Lough Neagh to the Donegal Mountains, having to northward Londonderry, and to southward Armagh, Monaghan, and Fermanagh. To the north and west it is hilly, the *Sperrin Mountains* being its chief heights, but as the hills sink towards the great lake on the east level tracts of cultivated land take their place, so that in this county agriculture is added to the linen trade. The principal town, *Omagh*, is in the centre of the shire; *Strabane* lies to the north of it in the Foyle valley; *Dungannon* has sprung up on a small coal-bed found to the eastward.

(6.) **Fermanagh** is only shut out from the Atlantic by a narrow belt of the Donegal Highlands. It lies between Tyrone on the north, and Leitrim and Cavan on the south, having Monaghan on its western side. The county is made up of the valley of the river Erne, which traverses its centre, widening into the two upper and lower lakes of the same name; while along the southern border a range of mountains forms the dividing wall between the Erne valley and the head-waters of the Shannon. In Fermanagh oats and wheat grow better than in the other northern counties, but among its little agricultural villages the only place of note is *Enniskillen*, which lies partly on an island formed by the Erne as it passes from lake to lake.

(7.) **Monaghan** is a small shire that stretches southwards from Tyrone, having Fermanagh and Cavan on its western border, while Armagh extends along its eastern side. Monaghan consists of a belt of high undulating ground with a gradual slope on one side to the basin of Lough Neagh and on the other to Lough Erne. Its two little towns lie one on each slope, *Monaghan* on a small river falling into Lough Neagh, and *Clones* on a stream which passes to Lough Erne.

(8.) **Armagh** is a narrow county which extends due north and south between Tyrone and Louth, and is bounded to west and east by Monaghan and Down. That part of the county nearest to the sea lies within the mountainous country of the coast; but as the hills slope to the Lough Neagh basin they form a fertile plain, in which are situated the larger towns of *Armagh*, a cathedral city, once the ecclesiastical capital of Ireland; and of *Lurgan* by Lough Neagh. Both these towns are famous for their manufactures of fine linen and damask.

(9.) **Cavan** is a hilly district driven down into the central plain of Ireland between Leinster and Connaught, and which thus forms the southernmost extremity of Ulster. On its northern side it is bounded by Fermanagh and Monaghan: Meath and Longford lie to the south, and Leitrim to the west.

The heights of Cavan lie at opposite ends of the county, rising to mountains in the north-west where

the Shannon takes its rise, and falling to low hills in
the south-east. Between these two tracts of highland runs the broad valley of the river Erne, with the
small lakes that are scattered along its course. The
land is not well cultivated, and the only town beyond
the size of a village is *Cavan.*

II. CONNAUGHT is in form a large irregular square
occupying the extreme west of Ireland, and running
out into the sea between the Bays of Galway and
Donegal. Its western part is very rugged and
mountainous, while the eastern passes into the
central plain of Ireland as far as the banks of
the Shannon, the change from the higher to the
lower ground being marked by a line drawn from
Galway to Sligo. The area of the province is 6,962
square miles, but its population is less dense than
that of any other part of Ireland; it contains in fact
under 847,000 inhabitants. It is peopled almost
entirely by the Celtic race, which is here hemmed in
between the sea and foreign invaders, and through
its remoteness from civilising influences has remained the wildest and least cultivated part of the
population of the island. The province is the least
fertile in Ireland; its farming is very poor, and its
chief industry lies in sheep-grazing. It has five
counties, of which four lie on the sea-coast.

(1.) **Leitrim**, the first of these shires, is a long
narrow district, just touching the sea-shore at Donegal
Bay. Lough Allen and the valley of the Shannon
divide the county into two parts; the northern part,
lying between the highlands of Fermanagh, Cavan,

and Sligo, is of mountainous character, while the southern part, stretching down to the plains of Longford and Roscommon, is low and flat. The whole county is poor, and does not possess a single town of any size. *Carrick-on-Shannon* is a mere village on its southern borders.

(2.) **Sligo** lies on the sea-coast between Leitrim and Mayo, with Roscommon to the south. A line of mountains runs along its north-eastern side, projecting into the sea: a second range of lower heights, the *Ox Mountains*, lies on its western border; and between these two ranges stretches a broad plain into which the sea has cut deeply so as to form the beautiful bay of Sligo, shut in by mountains on either side. On this bay lies the only town of any importance, *Sligo*, with 10,000 inhabitants.

(3.) **Mayo**, one of the largest counties in Ireland, borders the Atlantic Ocean, and is shut in to landward by Sligo, Roscommon, and Galway. The shire is made up of a mass of steep mountains rising precipitously out of the sea on the west, and bounded on the east by the broad valley of the river Moy, and by a tract of low ground belonging to the central plain of Ireland, in the northern part of which lies Lough Conn, and in the southern Lough Mask. The coast is wild in scenery, and deeply indented by bays, of which the chief are Killala, Blacksod Bay, landlocked by the peninsula of Mullet, and Clew Bay, with Clare Island and Achill Island near its mouth. To the north of Clew Bay rises the steep range of the Nephin Beg Mountains, and to the south of it the mountains of Croagh Patrick and Muilrea. The little towns of the shire lie in the low ground, such as *Castlebar, Ballina,* and *Killala* in the basin of the Moy, the latter at the head of Killala Bay; and *Westport* on Clew Bay.

(4.) **Galway,** the second largest county in Ireland, lies on Galway Bay, between the counties of Mayo and Clare. The greater part of it consists of an immense level plain extending far inland, and parted from the lowlands of Roscommon by the river Suck, from King's County by the Shannon, and from Tipperary by Lough Derg. But a smaller portion of the county is composed of a mass of heights, the Connemara Mountains, driven out seawards along the northern shores of Galway Bay, and almost cut off from the plain by Lough Mask and Lough Corrib. This rocky district is barren and deserted; and the few little towns of the shire lie in the river valleys of the plain. *Galway,* on the shores of the bay, is the largest town of Connaught, and occupies a position answering to that of Sligo, on its more northerly harbour; *Tuam,* the cathedral city, stands on the river Clare; and *Balinasloe,* on the Suck, is famous for the most important cattle fairs in Ireland.

(5.) **Roscommon** is the only inland county in Connaught. Its eastern boundary is formed by the Shannon and Lough Rea, which part it from Leitrim, Longford, Meath, and King's County; in the north and west it stretches out to Sligo and Mayo, while its southern part is narrowed by the Suck, which parts it from Galway. The county lies wholly within the central plain of Ireland, and forms an undulating tract of agricultural land, with a gentle slope from the higher ground of its northern borders towards the south. Its towns are scarcely more than villages : *Roscommon*, between the Suck and Lough Rea ; *Boyle*, between two small lakes in the north ; and near it *Elphin*, an episcopal town.

III. MUNSTER occupies the whole peninsula of south-western Ireland, from Galway Bay to Waterford Harbour, and forms the largest province in Ireland, having an area of 9,476 square miles. Its population amounts to 1,390,402 persons. It includes all the best natural harbours of the country, and two of its chief sea-ports, Cork and Limerick. The province comprises the great belt of the southern mountains ; but among these lie some of the most fertile districts in Ireland, such as the *Golden Vale* of Limerick and Tipperary, and the rich basin of the Shannon, districts whose fertility is increased by a climate milder and more full of moisture than that of any other part of Ireland. It has six counties, of which but one lies wholly inland.

(1.) **Clare** forms a peninsula, which is attached on the north-east to Galway, and bordered on all other sides by the Atlantic Ocean and the Shannon, the last of which divides it from Tipperary, Limerick, and Kerry. It is composed of a tract of rugged and mountainous country, broken across the centre by the valley of the Fergus, which contains the chief town, *Ennis*. *Kilrush* is a little port at the mouth of the

Shannon ; *Killaloe,* the seat of a bishopric, is situated on the Shannon as it issues from Lough Derg.

(2.) **Limerick** lies on the opposite side of the estuary of the Shannon from Clare, having Tipperary to the east, Cork to the south, and Kerry to the west. It is thus wholly shut out from the sea. A tract of hilly country runs along its western borders, and on the south-east rise the Galtees ; but the chief part of the shire consists of a broad plain of the richest land in Ireland, bordering the Shannon, and known from the extreme fertility of its soil as the *Golden Vale.* This district produces the best grass for cattle to be found in the United Kingdom. The county town, *Limerick,* is situated in the midst of the plain on the bank of the Shannon where it begins to widen to its estuary, and has become from the advantages of its position the largest town of western Ireland. It contains 40,000 inhabitants.

(3.) **Kerry,** which lies next to Limerick on the Shannon estuary, and thence projects into the Atlantic, forms the most south-western part of Ireland. To the south and east it is bounded by the county of Cork. Its coast is more rugged than any other part of the Irish shores, and thickly studded with small islands, of which Valentia is the chief. Two jutting peninsulas of rock stretch out seawards on its western side, between the deep bays worn by the Atlantic waves, the first ending in Mount Brandon, between Tralee Bay and Dingle Bay, and the second formed of the mass of the McGillicuddy's Reeks, between Dingle Bay and Kenmare Bay. The

McGillicuddy's Reeks end on their eastern side in the mountains and lakes of Killarney, well known for the beauty of their scenery. The industry of the county is small, and consists wholly of cattle-grazing, so that it has scarcely any towns save *Tralee*, a fishing-town on the bay ; and *Killarney*, whose chief livelihood depends on the tourists who visit the Lakes.

(4.) **Cork**, the largest county in Ireland, lies on its southern coast. The shire is composed of several long mountain ranges, which strike across it in parallel lines from Kerry to Tipperary, and which sink on its northern border into a fertile plain, forming part of the Golden Vale of Limerick. These ranges are parted by the valleys of three rivers running also from west to east, the Brandon, the Lee, and the Blackwater, in which the chief towns of the county lie sheltered. Each stream forms at its mouth an important harbour. Near the mouth of the Lee is the chief town, *Cork*, a large city with a cathedral, with woollen and linen manufactures, and with a considerable foreign trade due to its geographical position both with regard to Europe and to America. *Queenstown* is a port in an island of Cork harbour, and is important as the point of departure for America. The towns lying on the Bandon to the south of the Lee are *Kinsale* on the harbour at its mouth ; and *Bandon*, a little agricultural town higher up its valley. On the Blackwater are *Youghal*, on the harbour formed by the estuary of the river; and *Fermoy*, in its upper valley. Besides the river harbours there are considerable natural bays, such as *Bantry Bay*, and other lesser ones along the coast;

but these have as yet nothing but little fishing towns on their shores, being shut out from the central plain by barriers of rude mountains. The low lands of Cork form some of the best grazing country in Ireland, but the great moisture of the climate unfits them for the growing of wheat and corn.

(5.) **Waterford** is a small county lying next to Cork on the southern coast, and composed for the most part of the last heights of the same parallel mountain ranges which run across Cork. The Knockmeildown Mountains and the valley of the Suir part it on the north from Tipperary and Kilkenny. At its junction with the Barrow, the Suir forms the estuary of Waterford harbour, and near the head of this harbour is the large town of *Waterford*, the seat of the export trade to Bristol, with a population of 24,000. *Dungarvon* is a smaller seaport town on Dungarvon Harbour.

(6.) **Tipperary** is the only wholly inland county of Munster. It is shut in between eight other shires : to the east are King's County, Queen's County, and Kilkenny; to the south, Waterford; to the west the heights of the Galtees and Silvermine Mountains part it from Limerick and Clare, while Lough Derg and the Shannon sever it from Galway. The county practically consists of the basin of the Suir, a district which contains some of the best ground in Ireland for growing wheat, and whose farming is of a better kind than that of any of the counties we have passed through. The chief towns lie in this valley, and are wholly agricultural; *Thurles*, high up the river valley; *Cashel*, to the east of the Suir, built round the slopes of an abrupt rock, which is crowned by the ruins of one of the most famous ancient churches of Ireland, and still gives its name to a bishopric ; *Tipperary*, at some distance on the other side of the Suir; *Clonmel*, the county town, twice as large as any of these, with over 10,000 inhabitants ; and *Carrick-on-Suir*, on the border where the stream forsakes the county. *Nenagh* is a little agricultural town in the northern part of the shire where the land slopes down to Lough Derg.

IV. LEINSTER lies in the form of a long and rather narrow parallelogram, resting on the Irish Sea from the Mourne Mountains to the south-eastern point of the island. It consists for the most part of the great central plain with its low sea-border facing England, and was thus the first part of the country overrun by the English conquerors and parcelled out among English owners. It remains therefore the most English and least purely Celtic part of Ireland. The area of Leinster is 7,612 square miles; it contains 1,345,997 inhabitants. Of its twelve counties only five touch the sea-coast.

(1.) **Louth**, the northernmost of the Leinster counties, and the smallest shire in Ireland, extends along the low eastern coast between Carlingford Bay and the river Boyne ; while on its land side it touches the counties of Armagh, Monaghan, and Meath. The low ground which constitutes the bulk of Louth is mainly occupied by pasture lands; but in the harbours of the coast a considerable trade is carried on with England, which gives rise to two towns of some importance. These are *Dundalk*, on Dundalk Bay; and *Drogheda*, near the mouth of the Boyne.

(2.) **Meath** forms part of the central plain of Ireland, and only opens on the sea by a narrow belt of land running eastward between Louth and Dublin. To northward its broad reaches of pasture-land are bounded by the hills of Monaghan and Cavan; to westward they widen into the plains of Westmeath, King's County, and Queen's County; while to southward they pass into the more broken ground of Carlow. Within these limits stretches a level tract of exceedingly rich soil, which forms in fact the basin of the Boyne. Wholly given up to grazing industry, its towns are few and small. They all lie in the valley of the Boyne and of its affluent the Blackwater : *Trim*, on the Boyne; *Kells*, on the Blackwater; and *Navan*, at the junction of both rivers. The *Hill of Tara*, the ancient meeting-place of the Irish kings, is a few miles from Trim.

(3.) From Meath the county of **Westmeath** stretches westward between the shires of Longford and King's County. Lying mainly as it does within the basin of the Shannon, which flows along its western border, its rich tracts of level soil are watered by a number of small lakes and streams which drain their waters into that river. Its chief town, *Mullingar*, is situated in the centre of the county, in a district marked by a cluster of lakes and tarns; a second town, *Athlone*, is situated on the bank of the Shannon as the river passes out of Lough Rea.

(4.) The little county of **Longford,** which lies to northward of Westmeath, forms almost the central point of Ireland, where three of its provinces meet. On its northern borders Longford adjoins Cavan in Ulster ; and on its western side it is bounded by two counties of Connaught, Leitrim and Roscommon. Lying wholly in the basin of the Shannon, its surface is generally low and flat, and its pastures are watered by a multitude of small streams which pass westward to that river as it skirts the borders of the shire. On one of these small affluents lies the little county town, *Longford.*

(5.) The two following shires, those of King's County and Queen's County, were so named after Queen Mary and her husband, Philip II. of Spain, in whose reign they were first formed into counties. In the first of these the name of Philip is preserved in the village of Philipstown ; while the capital of the second, Maryborough, derives its name from Queen Mary.

King's County lies to the south of Westmeath and forms a long and narrow tract of land that extends westward to Roscommon and Galway, from which it is parted by the Shannon. To the south the Slieve Bloom Mountains divide it from Tipperary and Queen's County ; while to eastward it is bordered by Kildare. The great bulk of the shire lies like Westmeath in the basin of the Shannon, and is composed of the same rich grazing land. Its two most important towns, *Tullamore* and *Parsonstown,* lie to westward on affluents of the Shannon. In the eastern part of the county, however, the soil is less fertile, parts of it being composed of tracts of unreclaimed bog-land belonging to the Bog of Allen.

(6.) **Queen's County**, to the south of King's County, consists of a small circular tract of land of a mountainous character, bounded on the south by the hilly districts of Tipperary, Kilkenny, and Carlow. The Slieve Bloom Mountains run along its western border line, and in their heights lie the springs of the Barrow and the Nore. The towns are all agricultural, and are in fact little more than villages, such as *Maryborough* in its centre, and *Mountmellick* and *Portarlington* in the valley of the Barrow.

(7.) To westward of the King's and Queen's Counties, the shire of **Kildare** extends in a long tract of land due north and south, being bordered on its eastern side by Dublin and Wicklow. This shire consists for the most part of a level plain of pasture-land, watered in the south by the Barrow, and in the east by the Liffey, while its northern portion is occupied by a large part of the Bog of Allen. The towns of the county are very small. The chief among them are *Athy* on the Barrow in the south-west; *Naas*, and *Maynooth*, the great college for training Roman Catholic priests, on the *Liffey* in the east. Midway between Athy and Naas is the village of *Kildare*, an ancient episcopal city, and now the site of a military camp. A broad plain which surrounds the camp is known as the *Curragh of Kildare*, and forms a district famous for its sheep-pastures.

(8.) **Dublin** is a small narrow county, which lies on the eastern coast between the low plains of Meath and the mountains of Wicklow, having Kildare to the west. It is traversed by the lower valley of the river *Liffey*, that opens into the bay of Dublin, and whose banks, near the head of the bay, form the site of the capital of Ireland, DUBLIN. This city, the seat of an archbishopric, contains the chief University of Ireland ; its position at that opening of the central plain of Ireland which faces England makes it the natural centre of internal and foreign trade, and its population, 250,000, is consequently far greater than that of any other town in Ireland. Owing to the unsafe character of the bay on which it lies, its trade for the most part passes through *Kingstown*, a large town at the south-eastern opening of the bay, where a very fine artificial harbour has been made ; the north-eastern boundary of the bay is formed by the rocky promontory of Howth Head.

(9.) **Wicklow** extends southward from Dublin as far as Wexford, and is limited to westward by Kildare and Carlow. Though lying on the eastern sea-coast it differs from all the other maritime counties of Leinster, for while they form level plains Wicklow is composed of a rugged mass of granite mountains, second only in height to those of Kerry, which abruptly break the generally low shores of the Irish Sea. This circular mass of the Wicklow Mountains contains, as we have seen, some of the most beautiful Irish scenery ; rising to their greatest height in Lugnaquilla, they thence sink eastward to the sea,

and westward to the central plain. Three small
streams, which all alike have their sources in the
northernmost heights of the shire, are thrown down to
the Irish Sea, and are famous for the scenery along
their valleys, while each of them empties itself into
the sea by one of the chief towns of the county. The
Avoca, which has the longest course of the three, after
it has traversed the mountain group from north to
south and passed through the Vale of Avoca, forms
at its mouth the harbour where the seaport of *Arklow*
is built. The Vartry joins the sea near *Wicklow*,
the capital of the county, on Wicklow Harbour. The
Dargle flows past the base of the Great and Little
Sugarloaf Mountains to pour itself into the sea at
Bray, a fashionable watering-place.

(10.) To the south of Wicklow the county of
Wexford extends along the shore of St. George's
Channel, and forms the south-easternmost shire of
Ireland. It consists practically of the gradual slope
of the Arklow and Wicklow Mountains which lie along
its northern borders, and thence sink to the sea on its
southern and south-eastern limits ; the ground is therefore low and undulating for the most part. The river
Slaney passes through the centre of the county, forming
at its mouth the estuary of Wexford Harbour, with the
large trading town of *Wexford* on its inner side ; the
little county town of *Enniscorthy* lies farther up the
river valley. The river Barrow flows along a portion
of the western border of the shire, dividing it from
Kilkenny, and has on its banks the small town of
New Ross.

(11.) The hills which run along the western limits of Wicklow and Wexford divide both these shires from that of **Carlow**. This county consists of a very small tract of land which extends between the valleys of the Slaney and the Barrow; to the northward it widens to the plains of Kildare and Queen's County, while to westward it is shut in by the hills of Kilkenny. Its towns are small and of little consequence: *Carlow*, the county town, lies on the Barrow; and *Tullow* on the Slaney.

(12.) Across the western border of Carlow lies the much greater county of **Kilkenny**, which extends westward to Tipperary, and reaches north and south between Queen's County and Waterford. Lying in the basin of the Barrow, which bounds it to the east, the shire is watered by the two chief tributaries of this river, for its southern portion is bounded by the valley of the Suir, while the bulk of the county is composed of the valley of the Nore.

On the banks of this river, and in the centre of the rich plain which it traverses, lies *Kilkenny*, the capital of the county, and the largest inland town in Ireland. It contains about 17,000 inhabitants.

County.	County Town.	Area in Square Miles.	Population.
ULSTER—			
Donegal . . .	Lifford . . .	1,865	277,775
Londonderry .	Londonderry .	810	173,932
Antrim . . .	Carrickfergus .	1,190	419,782
Down . . .	Downpatrick .	957	217,992
Armagh . . .	Armagh . . .	513	179,221
Monaghan . .	Monaghan . .	500	112,785
Tyrone . . .	Omagh . . .	1,260	215,668
Fermanagh . .	Enniskillen . .	714	92,688
Cavan . . .	Cavan. . . .	746	140,555
		8,555	1,830,398
CONNAUGHT—			
Galway . . .	Galway . . .	2,447	248,257
Roscommon .	Roscommon .	950	141,248
Leitrim . . .	Carrick-on-Shannon . .	613	95,324
Sligo	Sligo	721	116,311
Mayo	Castlebar . .	2,231	245,855
		6,962	846,995
MUNSTER—			
Clare	Ennis. . . .	1,294	147,994
Limerick. . .	Limerick. . .	1,064	191,313
Kerry	Tralee . . .	1,853	196,014
Cork	Cork	2,885	516,046
Waterford . .	Waterford . .	721	122,825
Tipperary . .	Clonmel . . .	1,659	216,210
		9,476	1,390,402

County.	County Town.	Area in Square Miles.	Population.
LEINSTER—			
Longford . .	Longford . .	421	64,408
Westmeath . .	Mullingar . .	709	78,416
Meath . . .	Trim	906	94,480
Louth	Drogheda . .	315	84,199
Dublin . . .	Dublin . . .	348	405,625
Wicklow . . .	Wicklow . . .	781	78,509
Kildare . . .	Naas	654	84,198
King's County .	Tullamore . .	772	75,781
Queen's County	Maryborough .	664	77,071
Kilkenny . .	Kilkenny . .	795	109,302
Carlow . . .	Carlow . . .	346	51,472
Wexford . .	Wexford . .	901	132,536
		7,612	1,345,997

INDEX.

INDEX.

A.

ABBERLEY HILLS, 97, 98, 137
Aberdare, 239
Aberdeenshire, 330
Aberdeen, 331
Abergavenny, 143
Aberystwith, 234
Abingdon, 188
Accrington, 127
Achill Island, 365, 373, 384
Agriculture, 75
Agricultural counties, East Midland, 163, 165
Agricultural towns of Suffolk, 175; of Essex, 180; of Hertfordshire, 183; of Oxfordshire, 186; of Berkshire, 188; of Sussex, 197
Agriculture, of Wales, 218, 223; of Scotland, 283, 330; of Ireland, 381
Ague, 179
Airdree, 347
Aire, river, and its valley, 84
Alloa, 341
Almwch, 230
Alnwick, 117
Alyth, 336
Ambleside, 122
Andover, 198
Anglesea, Island of, 229
Angus, braes of, 265, 293
Annan, 355
Annan, river, 300
Annandale, 315
Anton, river, 94
Antrim, mountains of, 364
Antrim, 380
Appin, 200
Appleby, 122
Arbroath, 333
Arden, forest of, 140

Ardnamurchan Point, 308
Ardross, 263
Argyleshire, 318, 325
Arklow, 395
Armagh, county and city, 382
Arran, 310
Arran Islands, 373
Arran Mowddy Mountain, 222, 223
Arrenig Mountain, 222, 223
Arthur's Seat, 280
Arun, river, 94, 196
Ashton-under-Lyme, 128
Athlone, 391
Athol Mountains, 265, 335
Athy, 395
Atlantic Ocean, it effect on the climate of Great Britain and Ireland, 9, 27
Avon, river and valley (of Bristol), 98, 100, 107, 137, 139—141, 199—201, 208
Avon, river (of Salisbury), 94, 107, 139
Axe-edge Hill, 153
Aylesbury, 184
Aylesbury, vale of, 71, 184
Aylsham, 175
Ayrshire, 282
Ayr, county and town, 352, 353
Ayr, river, 282, 300

B.

BACUP, 127
Badenoch, 257
Bakewell, 149
Banbury, 186
Bandon, river, 370
Bala, lake, 222, 233
Ballina, 385
Ballanasloe, 385
Ballymena, 380

D D

Ballyshannon, 379
Balmoral, 293, 331
Bamborough Castle, 117
Bangor, 231
Banff, county and town, 329
Bann, river, 369
Bannockburn, 342
Bantry Bay, 374, 388
Barking, 181
Barmouth, 233
Barnard Castle, 119
Barnstaple, 212
Barra Islands, 310
Barrhead, 350
Barrow, river, 370
Barrowstoneness, or Bóness, 342
Barton-on-Humber, 166
Basingstoke, 198
Bassenthwaite Water, 45, 120
Bath, 208
Bathgate, 342
Bays, of the western coast of England
    and Wales, 30, 35, 39 ; of the eastern
    and southern coasts of England, 31,
    34, 35, 39 ; of Scotland, 303, 304,
    307, 308, 311, 322 ; of Ireland, 364,
    371, 373
Beachy Head, 34, 196
Beauley Firth, 290
Beaumaris, 230
Beaminster, 203
Beccles, 175
Bedfordshire, 163, 169
Bedford, 170
Belfast Lough, 374
Belfast, 380
Ben An. 335
Ben Attow, 257
Ben Avon, 264, 292
Benbecula, 310
Ben Cleugh, 280
Ben Cruachan, 258
Ben Dearg, 257, 265, 292, 335
Ben Hope, 257
Ben Klibreck, 262
Ben Lawers. 258, 335
Ben Lodi, 258
Ben Lomond, 258
Ben Mac Dhui, 264, 292
Ben More, 257, 258, 310, 335
Ben Nevis, 258
Ben Venue, 335
Ben Voirlich, 258
Ben Wyvis, 263
Ben-y-Gloe, 265, 335
Berkeley, 139
Berkshire, 179, 187
Berwickshire, 357
Berwick-on-Tweed, 117, 359

Berwyn Mountains, 47, 219, 221
Bideford Bay, 37, 212
Bideford, 212
Biggleswade, 170
Bill of Portland, 34
Birkenhead, 128
Birmingham, 141
Bishop Auckland, 118
Bishop Stortford, 183
Bishop Wearmouth, 118
Blackburn, 127
Black Country, 70
Blackdown Hills, 209
Black Larg, 273
Blacklead, 120
Black Mountains, 47, 48, 93, 219, 223
Blackpool, 125
Blacksod Bay, 373, 384
Blackwater, river, 180, 370
Blair Athol, 336
Blairgowrie, 336
Blyth, 117
Bochragh Hills, 366
Bodmin, 213
Bogs of Devon and Cornwall. 25, 211
Bog-land of Ireland, 367
Bolton, 127
Boot and shoe manufactures of Staf
    ford, 157 : of Northamptonshire
    168
Borrowdale, 45, 120
Boston, 166
Bournemouth, 199
Bowfell,, 43, 120
Boyne, river. 370
Bradford, Wiltshire, 202
Braemar, 293, 331
Braes of Angus, 265. 293
Braich-y-Pwll, 38, 220
Braintree, 180
Brampton, 120
Brandon Hill, 365
Bray Head, 375
Bray, Ireland, 395
Breadalbane Mountains, 258, 294
Brechin, 383
Brecknock Beacon, 47, 48, 99, 224
Brecknockshire, 235, 237
Brecon, 237
Breidden Hills, 236
Brendon Hills, 209
Brent, river, 181
Brentford, 183
Bridgnorth, 136
Bridgwater Bay, 37
Bridgwater, 209
Bridport, 203
Brighton, 197
Brine springs, 103

Bristol, 139
Bristol Channel and Harbour, 37
Bristol Channel, watershed and rivers
  of the, 95, 96, 226
Broad Law, 274
Broseley, 136
Brown Willy, 51, 213
Brue, river, 208
Buchan, 264
Buckinghamshire, 177, 178, 184
Buckingham, 185
Builth, 238
Bungay, 175
Bunnabeola, twelve pins of, 363
Bure, river, 174
Burnley, 147
Bury, 127
Bury St. Edmunds, 175
Buteshire; Bute, 326
Bute, 318
Bute; Kyles of Bute, 310
Buttermere, 45, 120
Buxton, 149

C.

CADER IDRIS, 55, 222, 233
Cahore Point, 375
Cairngorm, 264
Cairngorm Mountains, 264
Cairn Table, 280
Cairntoul, 264
Caithness, 262, 306, 318, 320
Calder, river, 126
Caledonian Canal, 254
Callander, 337
Cam, river, 171
Camborne, 213
Cambridgeshire, 163, 170
Cambridge, 171
Camel, river, 213
Campbell-town, 326
Campsie Fells, 280, 281
Canals, 74
Canals of the Fenland, 162
Cannock Chase, 69, 150
Canterbury, 192
Cantyre, 260
Cantyre, Mull of, 308
Cape Clear, 375
Cape Wrath, 259, 260, 289, 306
Paradise Hills, 135
Cardiff, 225, 239
Cardigan Bay, 36, 38
Cardiganshire, 234
Cardigan, 234
Carisbrook Castle, 200
Carlingford Bay, 374
Carlisle, 120

Carlow, county and town, 395
Carmarthen Bay, 37
Carmarthenshire, 240
Carmarthen, 241
Carnarvon Bay, 37
Carnarvonshire, 230
Carnarvon, 231
Carnedd Davydd, 220
Carnedd Llewellyn, 220
Carntogher Mountains, 364
Carrick, 282, 315, 353
Carrick-on-Shannon, 384
Carrick-on-Suir, 390
Carrickfergus, 380
Carse of Gowrie, 294
Cashel, 389
Castlebar, 385
Castle Douglas, 355
Castleton Caverns, 150
Catmoss, Vale of, 166
Cattle of Scotland, 330
Cavan, county and town, 382
Chalk ranges of uplands, 59, 172, 183, 198
Chalk cliffs of Kent, 192
Chalk downs of the Wealden, 73
Chalk downs, south, rivers of the, 91
Channel Islands; area and population, 18
Chard, 209
Charnwood Forest, 152
Chatham, 192
Chelmer, river, 180
Chelmsford, 180
Cheltenham, 139
Chepstow, 143
Cherry-orchards of Kent, 192
Cherwell, river, 92, 167, 186
Cheshire, 130; plain of, 70; rivers of
  the plain, 95, 101; pasture land, 103
Chesil Bank, 203
Chester-le-Street, 118
Cheviot Mountains, 10, 20, 21, 53, 270
Cheviot Peak, 275
Chichester Harbour, 196
Chichester, 197
Chiltern Hills, 25, 60, 65, 67, 91, 169, 183, 184, 186, 187
Chipping Norton, 186
Chorley, 127
Christchurch, 197
Cirencester, 140
Clackmannan, county and town, 339, 341
Clare Island, 373, 384
Clare, county, 386
Clay, 179, 191, 204

Clay, London, 181
Clear Island, 375
Clee Hills, 49, 135
Clent Hills, 97, 137
Cleveland Hills, 62, 158
Cleveland Moors, 155
Clew Bay, 373
Cliffs of the western coast of England and Wales, 30 : of the eastern and southern coasts, 31, 34
Climate of the British Islands, 3, 13, 27
Climate, affected by mountains, 40
Climate of Wales, 218
Clitheroe, 127
Clones, 382
Clonmel, 390
Clun Forest, 135
Clwyd, river, 101, 221, 226, 231
Clyde, river, 283, 299
Clyde, Firth of, 260, 307
Clyde, counties of the, 345
Clydesdale, 315, 347
Coal-measures of Great Britain, 6
Coal, 12, 16, 17, 27, 42, 48, 70, 85, 138
Coal-fields; Ashby-de-la-Zouch, 153 ; Cumberland, 121; Derbyshire, 149; Durham, 118 ; Northumberland, 117 ; Pennine Moors, 53; Shropshire, 136 ; South Lancashire, 126 ; Staffordshire, 151 ; Wales, 143, 144, 218, 224, 232, 239
Coal-fields of England and Wales, tabular statement of their area, 112, 113
Coal-fields of Scotland, 11, 248, 282, 284, 296
Coalbrook Dale, 136
Coast-lines, importance of, 29
Coast-line of England and Wales, 21, 29, 32; of Scotland, 302; of Ireland, 370
Coatbridge, 347
Cockermouth, 121
Coilin Mountains, 309
Colchester, 180
Colne, 127
Colne, river, 92, 180, 181, 183, 184
Colonization, 17
Colonsay Island, 310
Commerce, result of geographical position on, 7, 40
Commerce of England and Wales, 75, 82, 88, 90—92, 95, 182 ; of Scotland, 248, 284, 286, 291, 294, 300; of Ireland, 7, 382
Congleton Edge, 130
Coleraine, 379
Coniston Old Man, 44, 55

Coniston Lake, 129
Coniston Water, 45
Connaught, 377, 383
Connaught, mountains of, 365
Connemara Lakes, 368
Connemara Mountains, 365
Conway, river, 101, 221, 226, 231
Conway, 231
Copper-mines, 52, 213
Copper foundries of Wales, 224, 225
Cork Harbour, 370, 375
Cork, county and town, 388
Cornsare Point, 363
Cornwall, 205, 206, 212
Cornwall, Highlands of, 25, 41, 49, 51, 78
Cornwall, rivers of, 94
Cotswold Hills, the, 23, 61, 78, 91, 97, 138, 185
Cotton manufacture, 42, 104, 127, 150
Counties, English, 108 ; of the Ribble and Mersey basins, 123; of the Severn Basin or West Midlands, 133 ; of the Humber Basin, 145; of the Trent Basin, 146, 147 ; of the Wash or East Midland Counties, 162, 164; East Anglian, 172, 173; of the Thames Basin, 177, 178 ; of the Southern Waterparting, 194 South-western, 205, 206
Counties of England ; tabular view of areas and population, 214
Counties, Welsh, 228, 229 ; of the Irish Sea Basin and the Bristol Channel Basin ; table of areas and population, 242
Counties of Scotland, 315, 317, 351; table of areas and population, 360
Counties of Ireland, 376; table of area and population, 397
County towns, 109
Coventry, 141
Cowal, 260
Cowes, 230
Cranborne Chase, 202
Craven Moors, 156, 159
Crewkerne, 208
Cricklade, 201
Crieff, 336
Criffel, 273
Croagh Patrick, 365
Cromarty Firth, 290, 305
Cromarty, plain of, 263
Cromarty, Ross and, 318
Cromarty, county and town, 321, 322
Cross Fell, 41, 54, 114
Crouch, river, 180
Croydon, 189
Crummock Water, 45, 120

Culver Cliffs, 199
Cumberland, the county, 119; its hills
   and lakes, 24, 43, 44
Cumbrays, the, 311
Cumbrian Hills, 24, 41, 43, 44, 54,
   113, 119
Cumbrian rivers, 96, 105
Cunninghame, 282, 352
Cupar Angus, 336, 340
Curragh of Kildare, 395

D.

DALWINNIE, Pass of, 265
Dalkeith, 344
Darent, river, 191
Darlington, 119
Dart, river, 211
Dartmoor, 50, 78, 210, 212
Dartmouth, 212
Darwen, river and valley, 126, 127
Daventry, 168
Deal, 192
Dean Forest, 49, 138
Dee, river and estuary, 46, 78, 100,
   101, 107, 130, 221, 226, 231, 233
Dee, river and valley, Scotland, 264,
   285, 291, 292, 300
Denbighshire, 231
Denbigh Hills, 221
Denbigh, 231
Deptford, 192
Derbyshire; the Peak, 22; the Pen-
   nine chain of hills, 23, 41
Derbyshire, coal measures of, 119
Derwent, river, 45, 84, 86, 105, 118,
   120, 149, 156
Derwentwater, 45
Devizes, 201
Devonport, 211
Devonshire hills, 25, 41, 49, 50
Devonshire rivers, 94
Devonshire, 205, 206, 210
Dingle Bay, 374
Dingwall, 322
Dirie More, 257
Diss, 175
Dolgelly, 233
Dollar Law, 274
Don, river, 85, 156, 291, 292
Donegal, Mountains of, 364
Donegal, Bay of, 373
Donegal, county and town, 378, 379
Dorchester, 203
Dorking, 189
Dornoch Firth, 290, 305, 322
Dornoch, 321
Dorset Heights, 64, 202

Dorsetshire, 194, 195, 202
Doune, 337
Dove, river, 86, 149, 150
Dover, Straits of, 26, 34, 192
Dover, 192
Doveran, river, 291
Dovey, river, 222, 226, 233
Dowlais, 239
Down, county, 380
Downpatrick, 381
Downs (see Cotswold Downs, Hamp-
   shire Downs, Marlborough Downs,
   North Downs, South Downs)
Drainage of the Fens, 71
Drainage, 179
Drogheda, 391
Droitwich, 137
"Drumalban," the backbone of Scot-
   land, 256
Dryburgh Abbey, 298
Dublin Bay, 364, 374
Dublin, county and city, 394
Dudley Hills, 69
Dudley, 138
Duddon river, 105, 120
Dumbarton, county and town, 349
Dumfries, county and town, 352, 355
Dunbar, 345
Dunblane, 337
Duncansby Head, 306
Dundalk Bay, 374
Dundalk, 391
Dundee, 333
Dundrum Bay, 374
Dunfermline, 340
Dungannon, 381
Dungarvon, 389
Dunge Ness, 34
Dunkeld, 336
Dunkerry Beacon, 50, 209
Dunmore Head, 373
Dunnet Head, 306
Dunoon, 326
Dunse, 359
Dunstable, 170
Durham, plains of, 114
Durham, county, 117
Durham, the city, cathedral, and
   castle, 118
Durnley, 139
Dysart, 350

E.

FARN, river, 295
East Anglian counties, 172, 173
East Anglian heights and uplands, 25,
   33, 60, 65, 67, 89, 91, 163, 172, 174,
   175

Eastbourne, 197.
Eastern counties of England, 110, 111
East Midland counties, 162, 164
Eastern highlands of Scotland, 327
East Swale, river, 93
Eden, river, 43, 105, 107, 115, 120, 122
Edge Hills, 61, 140, 186
Edinburgh Rock, 280
Edinburgh, county and city, 284, 296, 343, 344
Egremont, 121
Elgin, county and town, 329
Ellesmere, 136
Ely, Isle and city of, 170
Emerald Isle, the, 367
England, its area, 12 ; its characteristics compared with those of Wales, Scotland, and Ireland, 13 ; its position in relation to the continent, 13 ; effect of structure and position on its political history, 14 ; its area and population, 18 ; view of Western England, 23 ; view of Eastern England, 25 ; west coast, 35
England and Wales ; general view of, 19 ; their historical and political differences, 27
English Channel, its effect on the civilisation of Britain, 15, 20, 32, 34
English counties, 108, 114
Ennerdale Water, 45
Ennis, 386
Enniscorthy, 395
Enniskillen, 382
Epping Forest, 72, 179
Epsom, 189
Erne, river and lakes, 368, 369
Errigal, Mount, 364
Erris Head, 363, 373
Esk, river, 300, 301 (see North and South Esk)
Eskdale, 315
Essex, 177—179
Estuaries of the English Coast, east, south, and west, 33, 36, 39
Estuaries : Devonshire rivers, 211 ; Irish rivers, 369, 375 ; Medway, 192 ; Mersey, 103, 104 ; Clyde, 283, 284, 299 ; Forth, 283, 284 ; Severn, 93, 139, 226 ; Thames, 72, 90, 131
Eton, 184
Ettrick Forest, 315
Ettrick Pen, 275
Ettrick, river, 298
Europe ; its relation to the British Islands, 1
Evenlode, river, 186
Evesham, Vale of, 137

Evesham, 138
Exe, river, 50, 94, 210, 212
Exeter, 212
Exmoor, 78, 209, 210
Eye, 175

F.

FAIR HEAD, 363, 372
Fal, river, 213
Falkirk, 342
Falmouth, 213
Faringdon, 188
Farne Islands, 117
Farnham, 189
Fenlands, 71, 88, 89, 162, 165
Fermanagh, county, 381
Fermoy, 388
Fever, 179
Ffestiniog, 222, 223
Fifeshire, 339
Fife, Hills of, 280
Findhorn, river, 291
Firths of Scotch rivers, 287, 260, 299, 307 ; Forth, 395 ness, 267, 305 ; Lorn, 307 ; 307
Fisheries of England, 175, Scotland, 248, 307, 326
Flamborough Head, 33
Flamborough Hills, 62
Fleetwood, 125
Flintshire, 232
Flint, 232
Flint Hills, 221
Folkestone, 191, 192
Forelands, North and South,
Forests ; Epping and Hains 199 ; of Arden, 140 ; the Ne 74 ; of Wyre, 135
Forest Ridge, the, 73
Forres, 329
Forth, firth of, 395
Forth, river, 283, 295, 305
Fort William, 324
Foulness Island, 33
Foyle, river, 369
Fraserburgh, 330
Frome, river, 94, 203
Frome, 208
Fuller's earth, 190
Furness, peninsula of, 44, 129

G.

GAINSBOROUGH, 165
Gala, river, 298

Galashiels, 298, 358
Galloway, 315
Galloway Hills, 273
Galloway, Mull and Rhinns of, 308
Galtees Hills, 366
Galway Bay, 364, 373
Galway, county and town, 385
Gateshead, 119
Geography, Political (see Political Geography)
Giant's Causeway, 372
Glamorganshire, 238
Glamorgan, Vale of, 240
Glanford Brigg, 166
Glasgow, 284, 299, 347
Glas Mhiel, 265
Glenelg, 260
Glengariff, 365
Glenmore, 254, 262, 263
Glens of Scotland, 260, 261, 286
Glossop, 150
Gloucester, county and city, 133, 138, 139
Glyders, the, 220
Gneiss of the Scottish Rocks, 253, 260
Goat Fell, 310
Golden Vale of Limerick, 386, 387
Gosport, 199
Govan, 347
Gourock, 350
Gowrie, Carse of, 294
Grampian Mountains, 255
Granite quarries of Cornwall, 52; of Scotland, 254
Grantham, 166
Gravesend, 192
Great Britain; general characteristics and position of, 1; its area and population, 18
Great Britain and Ireland, their political and social differences, 8; political division of Great Britain, 9
Great Marlow, 184
Great Orme's Head, 37, 220
Great Ouse, river, 80, 107, 169, 170, 174
Greenlaw, 359
Greenock, 350
Greenwich, 192
Grimsby, 166
Guildford, 189

H.

HADDINGTON, county and town, 345
Hadleigh, 176
Hainault Forest, 72, 173
Halstead, 180

Hambledon Hills, 62, 158
Hamilton, 347
Hampshire, 194, 195, 197
Hampshire Downs, 65, 67, 78, 187, 189, 194, 197
Hampstead, 181
Harbours of England and Wales, 14, 29, 30, 33, 76, 91, 93, 94, 105, 128, 173, 175, 196, 199, 203, 224, 230; of the South of England, 35; of the West Coast, 36; of Ireland, 370, 374, 575
Harlech, 233
Hartlepool, 119
Harris, island of, 310
Harrow, 181, 183
Hart Fell, 274
Hartland Point, 38
Harwich, 180
Haslingden, 127
Hastings, 197
Hawes Water, 45
Hawick, 358
Hay, Radnorshire, 238
Headlands of England, East, South, and West, 33, 34, 37, 38; of Scotland, 306, 308, 311; of Ireland, 363
Heather in Scotland, 257
Hebrides, the, 309
Helensburgh, 349
Helvellyn, 43—45, 55, 121
Henley, 186
Hereford, Plain of, 47, 48
Herefordshire, the county, 133, 142
Hereford, 142
Hertfordshire, 177, 178, 183
Hertford, 183
Hexham, 117
Heywood, 128
High Dyke, Lincolnshire, 165
Highgate, 181
Highlands of Scotland, 10, 247, 251, 252; contrasted with the Lowland Hills, 270
Highland Lakes of Scotland, 266
Highland Mountains of Scotland, 276
Highland Rivers of Scotland, 289; tabular view of them, 301
Highlands of Ireland, Northern, 364; Western, 365; Southern, 365; Eastern, 366
High Wycombe, 184
Hitchin, 184
Hog's Back, 189
Holderness Marsh, 62
Holland, or the hollow land, Lincolnshire, 71, 165
Holywell, 232

Holyhead and Holyhead Island, 38
230
Holy Island, 33, 117, 230
Honiton, 212
Hop-grounds of Kent, 192
Horncastle, 166
Horn Head, 372
Horsham, 197
Hounslow, 183
Howth Head, 374
Hull, port of, 87
Humber Basin, counties of the, 145
Humber river-group, 83
Humber, the river, 33, 104, 107, 165;
its estuary, 80, 83, 87
Hungerford, 188
Hunstanton Point, 33
Huntingdonshire, 163, 168
Huntingdon, 169

I.

ILFRACOMBE, 210
Ilsley Downs, 65, 91, 187
Ingleborough, 41
Inverary, 326
Inner Hebrides (see Hebrides)
Inverness, Firth of, 267, 290, 305
Invernesshire, 318
Inverness, county and town of, 323, 324
Ipswich, 175
Ireland; general characteristics and
position of, 1, 7; its political and
social differences from Great Britain,
8; its area and population, 18
Ireland, 362
Irish Sea, 20, 21, 25, 35, 102
Iron and ironworks in England, 12, 42,
48, 70, 82, 121, 129, 131, 136, 138,
144, 149, 224, 232, 240; in Scotland,
248, 283, 284, 296
Ironbridge, 136
Irvine, 353
Irwell, river, 103, 126, 128
Isla, river, 295
Islay, Island and Sound of, 310
Islands on the coasts of England, 33,
35, 38
Island of Anglesea, 35, 229
Isle of Man; its area and population,
18, 243
Isle of Portland, 203
Isle of Sheppey, 191, 193
Isle of Skye, 309
Isle of Thanet, 191
Isle of Wight, 34, 199
Islands of Scotland, 302, 304, 308,
312

Islands of Ireland, 370, 372, 373
Isle, river, 209
Itchen, river, 94, 198
Ivel, river, 209

J.

JEDBURGH, 359
Johnstone, 350
Jura; Pass of Jura, 310

K,

KELLS, Rhinns of, 273
Kells, 391
Kelso Abbey, 298
Kelso, 359
Kendal, 122
Kenilworth Castle, 141
Kenmore Bay, 374
Kennet, river, 92, 187, 201
Kent, 179, 190
Kent, river, 122
Kerry, county, 387
Kerry Head, 373
Kerry Mountains, 365
Keswick, 120
Kettering, 168
Kidderminster, 137
Kildare, county and town, 395; the
Curragh, 395
Kilkenny, county and town, 396
Killala, 385
Killala Bay, 373, 384
Killaloe, 386
Killarney, 388
Killarney Lakes, 368
Killiecrankie, Pass of, 265
Kilmarnock, 353
Kilrush, 386
Kincardineshire, 332
King's County, 392
Kingston-on-Thames, 190
Kingston, Hampshire, 199
Kingstown, 374
Kingstown, Ireland, 394
Kingussie, 324
Kinross, county and town, 339, 340
Kinsale Harbour, 370, 375
Kinsale, 388
Kinsale, Old Head of, 375
King's Lynn, 174
Kirkaldy, 340
Kirkcudbright, county and town, 352,
354
Kirkham, 125
Kirkintilloch, 349

Kirkwall, 319
Knapdale, 260, 308
Knighton, 237
Knockmealdown Hills, 366
Kyle, 353

L.

LAGAN, river, 370
Lagan, 380
Lakes ; of Cumberland, 24, 44, 120 ; of the Highlands of Scotland, 266 ; of Southern Scotland, 275 ; of Perthshire, 335 ; tabular view of Scotch Lakes, 277 ; of Ireland, 9, 367
Lambeth, 190
Lammermuir Hills, 270, 274
Lanark, county and town, 346, 347
Lancashire ; manufactures of, 42 ; rivers of the plain of, 95, 101 ; the county, 123
Lancaster, 125
Land's End, 34, 35, 38, 51
Language, 17
Lauder, river, 298
Lauderdale, 315
Lea, river, 92, 179, 181, 183
Lead mines, 52, 149, 213, 232
Leam, river, 141
Leamington, 141
Ledbury, 142
Lee, river, 370
Leigh, 128
Leighton Buzzard, 170
Leinster, 377, 390
Leith, 344
Leitrim, county, 383
Lennox, 315
Leominster, 142
Lerwick, 319
Letterkenny, 379
Leven, river, 300
Lewes, Sussex, 197
Lewis, Island of, 310, 322
Lickey Hills, 97, 137
Liddel, river, 301
Liddesdale, 315
Liffey, river, 370
Lifford, 379
Limerick, county and town, 387
Linen manufactures in Scotland, 242, 294, 296 ; in Ireland, 7, 382
Lincoln Wolds, 61, 83, 165
Lincolnshire, 163
Lincoln, 166
Linlithgow, county and town, 343
Lizard Head, 35
Lizard Point, 51
Llanberis, Pass of, 219

Llandaff, 239
Llandeilo, 241
Llandovery, 241
Llandudno, 231
Llanelly, 241
Llangollen, Vale of, 231
Llanidloes, 236
Loch Alsh, 307
Loch Arkaig, 267
Loch Awe, 368
Loch Carron, 307
Loch Earn, 268, 335
Loch Edderachillis, 307
Loch Eil, 267, 307
Loch Enard, 307
Loch Ericht, 267, 335
Loch Etive, 268, 307
Loch Fyne, 268, 308
Loch Garry, 267
Loch Hourn, 307
Loch Katrine, 268, 335
Loch Ken, 275
Loch Laggan, 267
Loch Leven, 268, 307
Loch Linnhe, 307
Loch Lochie, 267
Loch Lomond, 258, 259, 268, 280
Loch Long, 268, 308
Loch Morar, 267
Loch-na-gar, 265
Loch Ness, 267
Loch Oich, 267
Loch Quoich, 267
Loch Rannock, 267, 335
Loch Shiel, 267
Loch Shin, 266
Loch Spey, 290
Loch Sunart, 307
Loch Tay, 267, 335
Loch Torridon, 307
Lochaber, 257
Lomond Hills, 280
London ; its position as affecting commerce, 7, 72, 92, 181 ; its population, 182
Londonderry, county and town, 379
Longford, county and town, 391
Long Mynd, 49, 135
Longtown, 120
Loop Head, 373
Lothians ; East, Middle, and West, 315
Lothians, the, 343
Lorn, 260
Lorn, Firth of, 307
Lough Allan, 368
Lough Conn, 361
Lough Corrib, 368
Lough Derg, 368

Lough Foyle, 372
Lough Mask, 368. 384
Lough Neagh, 368
Lough Rea, 368
Lough Swilly, 372
Louth, 166
Louth, county, 390
Lowestoft, 175
Lowlands of Scotland, 10, 247; their hills and plain, 248, 269, 276; contrasted with the Highlands, 270; grouping of the hills, 272; their rivers, 289; tabular view of them, 301
Lowther Hills, 273
Luce Bay, 308
Ludlow, 136
Lug, river, 142
Lugnaquilla, 366
Lune, river, 43, 105, 115, 122, 125
Lurgan, 382
Luton, 170
Lytham, 125

M.

MACCLESFIELD FOREST. 130
Macgillicuddy Reeks, 365
Madeley, 136
Maidenhead, 188
Maidstone, 191
Mainland, Orkney Islands, 319
Maldon (Essex), 180
Malin Head, 372
Malmesbury, 202
Malvern Hills, 49, 97, 98, 137
Malvern, 138
Man, Isle of (see Isle of Man)
Manchester, 128
Manufactures of England and Wales, 42, 69, 70, 75, 85, 87, 103, 104, 112, 119, 127, 129, 131, 132, 136—139, 141, 149—151, 153, 154, 160, 166, 168, 173, 175, 185, 212, 236; of Scotland, 248, 249, 298, 331, 348, 350, 358: of Ireland, 7, 382
Marble quarries, 211
March (Cambridgeshire), 171
Margate, 192
Market Drayton, 136
Marlborough Downs, 65, 66, 91, 187, 201
Marlborough, 201
Marsh lands and Marshes, 5, 88, 165, 179, 191, 207
Maryborough, 393
Maryport, 121
Matlock, 149
Maynooth, 395

Mayo, Mountains of, 365
Mayo, county. 384
Mealfourvoonie, 263
Meath, county, 391
Medway, river, 92, 94, 191
Melrose Abbey, 298
Melrose, 359
Menai Strait, 35
Merionethshire, 233
Merrick Mountain, Scotland, 273
Mersey, river, 36, 101, 103, 107, 125, 130
Mersey Basin, counties of the, 123
Merthyr Tydvil, 239
Metallic products and metals, 6, 40, 43, 50, 117
Mickle Fell, 41, 54
Middlesex, 117, 178, 181
Middleton, 128
Midhurst, 197
Midland Counties of England, 110, 111
Milford Haven, town and harbour of, 37, 241
Minch, straits of, 310
Mines and minerals of England, 6, 12, 13, 16, 27, 40, 50, 69, 112, 117, 121, 127, 211, 213, 224; of Wales, 224, 225; of Scotland, 296; in the Isle·of Man, 243
Mineral springs of Derbyshire, 149; of Leamington, 141; of Malvern, 138; of Reigate, 189
Mizen Head, 363
Moel Siabod, 220
Moffat, river, 300
Moffatdale, 356
Mold, 232
Mole, river, 92, 94, 189
Monadh-Leadh Mountains. 264
Monaghan, county and town, 382
Monk Wearmouth, 118
Monmouthshire, 133, 143
Monmouth, 143
Monnow, river, 142
Montgomeryshire, 235
Montgomery, 236
Montrose, 333
Moorlands and Moors of England, 25, 42, 53, 83—85, 114, 126, 159, 209, 213; of Wales, 48; of Scotland, 247, 257, 258, 262
Moray (Elgin), 329
Moray Firth. 261, 305, 322; rivers of the Firth, 290
Morpeth, 117
Morvern, 260
Morvern Hills, 262
Morecambe, Bay, 36, 43

Motherwell, 347
Mounta.ns, importance of, 40
Mountain groups of England and Wales, 40
Mountains of England, 43, 51, 105, 114, 115, 121, 135, 143, 205, 209, 213, 217; of Wales, 11, 45, 49, 218, 243; of Scotland, 10, 246—248, 251, 254, 257, 258, 265, 321, 328; of the Scotch Highlands and Lowland Hills, tabular view of, 276; of Ireland, 363
Mounta.n torrents of Scotland, 286
Mountmellick, 393
Mount's Bay, 35
Mourne Mountains, 366, 374
Mowcop, 150
Moy, river, 369
Much Wenlock, 136
Mulrea, 365
Muirfoot, Hills, 274
Mull, Island and Sound of, 309, 310
Mull of Cantyre, 308
Mull of Galloway, 308
Mullet, peninsula of, 373, 384
Mullingar, 391
Munster, 377, 386
Musselburgh, 344

N.

Naas, 395
Nairn, county, town, and river, 291, 328
Naval power of Great Britain, 16
Navan, 391
Neath, river, 224, 227
Neath, 240
Needles, the, 199
Nen, river, 89, 107, 167, 168, 170
Nenagh, 390
Nephin Beg Mountains, 365
Ness, river, 290
Newark-on-Trent, 154
Newbury, 188
Newcastle-on-Tyne, 117
New Forest, 74, 179, 199
Newmarket, 171
Newport, Isle of Wight, 200
Newport, Monmouthshire, 143
Newport, Shropshire, 136
New Radnor, 237
New Ross, 395
Newry, 381
Newton, 128
Newton Abbot, 212
Newtown, Montgomeryshire, 236
Newtown Ards, 381
Newtown Limavady, 379

Nidd, river and valley, 84
Nith, river, 272, 300
Nithsdale, 315
Nore, river, 370
Norfolk, 172—174
Northern counties of England, 110, 111
North-western counties of England, 110, 111
Northampton Heights, 61
Northampton Uplands, 83
Northamptonshire, 163, 167
Northampton, 168
North Berwick Law, 280
North Downs, 26, 60, 66, 67, 93, 189—191
North Esk, 293
North Foreland, 34
North Sea, its effect on the civilisation of Britain, 15, 20, 21, 25, 78; its watershed, 81
North Shields, 117
Northumbrian counties, 114, 115
Northumbrian rivers, 80
Northumberland, 116
North Uist, 310
North Walsham, 175
Norwich, 176
Noss Head, 306
Nuneaton, 141

O.

Oakham, 166
Oban, 326
Ock, river and valley, 187, 188
Ochill Hills, 280, 281, 294
Oldbury, 138
Oldham, 128
Old Man Mountain, 129
Olney, 185
Omagh, 381
Oolitic range of Uplands, 15, 52, 61, 83, 88, 163, 167, 185, 200, 202, 203
Ord of Caithness, 262, 306
Orkney Islands, 308, 309
Orkney, 318, 319
Oronsay Island, 310
Orwell, river, 93, 175
Oswestry, 136
Ouse, valley, 63, 69, 71
Ouse, river, and its tributaries, 83, 84, 87, 94, 107; its basin, 143, 154, 167, 168, 170, 185, 190
Outer Hebrides (see Hebrides)
Over Darwen, 127
Oxfordshire, 177, 178, 185
Oxford, 186
Ox Mountains, 384

P.

PADIHAM, 127
Paisley, 350
Parret, river, 49, 95, 100, 207, 209
Parsonstown, 392
Pass of Dalwinnie, 265
Pass of Killiecrankie, 265
Pass of Llanberis, 219
Pasture-lands, of England, 103, 209; of the Scotch mountains, 248; of Ireland, 367, 386
Patrick, 347
Peak of Derbyshire, 22, 41, 55, 86
Peal Fell, 275
Peebles, county and town, 356, 357
Pembrokeshire, 241
Pendle Hill, 125
Pennine Chain of Moors and Mountains, 22, 23, 41, 52, 59, 69, 79, 83, 156; its coal-fields, 112
Penrhyn slate quarries, 220
Penrhyn, 231
Penrith, 120
Pentland Hills, 280
Penzance, 213
Perry, river, 135
Perth, county and town, 294, 331, 334, 336
Peterborough, 168
Peterhead, 330
Petersfield, 198
Pewsey, Vale of, 201
Physical structure and resemblance of the British Islands, 2: their differences, 3, 6
Pickering, Vale of, 62, 84, 155, 157
Pitlochrie, 336
Plains of England, 23, 62, 68, 83, 84, 114, 162, 194; of Scotland, 263, 278, 305, 338; of Ireland, 363, 366, 377
Plinlimmon Range of Mountains, 47, 219
Plinlimmon, 55, 222
Plym, river, 211
Plymouth Sound, 35, 95, 211
Plymouth, 211
Polden Hills, 209
Political geography, 8; political and social differences of Ireland and Great Britain, 8; the political division of Great Britain, 9; of Wales, 11; of Scotland, 10, 313; of England, 12; of the English counties, 108, 110, 160
Political history of England affected by its structure and position, 14
Pollockshaws, 350
Pomona, 319

Poole Harbour, 94, 203
Poole, 204
Population of Great Britain generally, 16; of London, and Scotland, 16, 17; of Lancashire, 129; of Scotland, 'Gaelic and English, 10 (see statistics under each County)
Portarlington, 393
Port Glasgow, 350
Portland Bill, 34
Portland, Isle of, 203
Portobello, 344
Portpatrick, 269, 272, 353
Portree, 325
Ports of England, 197; principal, Bristol, 98, 139; Falmouth, 213; Glasgow, 299; Grimsby, 166; Hull, 87; Liverpool and Birkenhead, 104, 128; London, 182; Newcastle, 82; Sunderland, 82; Poole, 204: Yarmouth, 174; Cardiff, 225; of Scotland, 284, 296; of Ireland, 372, 378, 388
Portsea, 199
Portsmouth and its harbour, 35, 74, 199
Position of the British Islands, 7
Potteries, the, 151
Presteign, 237
Preston and its harbour, 105, 125
Provinces of Ireland, 277
Purbeck Island and its Cliff-line, 74, 203
Purbeck Heights, 203

Q.

QUANTOCK HILLS, 50, 209
Queen's County, 392, 393
Queenstown, 388

R.

RADNORSHIRE, 235, 236
Radnor Forest, 237
Railroads, 74
Rainfall of England, 4, 13, 27, 76; of Wales, 218
Ramsgate, 192
Rannoch, Moor of, 257, 258
Rathlin Island, 372
Reading, 188
Redditch, 138
Redruth, 213
Reigate, 189
Renfrew Hills, 280
Renfrew, county and town, 349, 350
Rhinns of Galloway, 308
Rhinns of Kells, 273

Rhyl, 233
Ribble, river, 101, 104, 125, 156
Ribble Basin, Counties of the, 123
Richmond, Surrey, 190
Richmond Moors, Yorkshire, 155, 159
Ridings of Yorkshire, 160
Rivers, 5
Rivers of England and Wales, 20, 49, 51, 73, 74
River system of England, 13, 75; determined by mountains, 40
Rivers, importance of, 75
Rivers, grouping of, 79
River groups, table of, 106
River system of Wales, 225; of Scotland, 285; of Ireland, 368
Roads, 20, 74
Rochdale, 127
Rochester, 191
Rocks of England, 6, 31, 47; of Wales, 216, 220; of Scotland, 246, 252, 270
Rockingham Forest, 61, 89
Roden, river, 135
Roding, river, 180
Romsey Marsh, 191
Ross, Herefordshire, 142
Ross and Cromarty, 318, 321
Roscommon, county and town, 385, 386
Rother, river, 191
Rothesay, 327
Rough Tor, 51
Roxburgh, 357, 358
Ruabon, 221, 232
Rugby, 141
Rutherglen, 347
Ruthin, 231
Rutlandshire, 163, 166
Ryde, 200

S.

Saffron, 180
Saffron Walden, 180
St. Abb's Head, 269, 272
St. Alban's, 184
St. Alban's Head, 34
St. Andrew's, 340
St. Asaph, 232
St. Bee's Head, 37
St. Bride's Bay, 37, 241
St. David's Head, 38
St. George's Channel, 35
St. Helena, 198
St. Ives, Huntingdonshire, 169
St. Ives, Cornwall, 213
St. Leonard's, 197

St. Mary's Loch, 275
St. Neots, 169
St. Ninian's, 342
Salford, 128
Salisbury Plain, 64, 66, 78, 194
Salt mines, 13, 103, 118, 130, 137
Salt manufacture, 119
Scafell, 43, 55, 120
Scenery of the Highlands and Lowlands of Scotland, 271, 292, 303; of Ireland, 368
Schiehallion, 258, 335
Scilly Islands, 34, 51
Scone, 337
Scotland; political geography of, 10; Highlands and Lowlands, 10; Clans, 10; progress of civilisation in, 11; its area, 11; union with England, 11; coal-beds, 11; manufactures, 11; compared with Wales, 12; its area and population, 18, 246, 249; physical map of, 244; boundaries, 245; physical structure, 246; industries, 248; coast-line and islands, 302; political divisions, 313
Sea-lochs of Scotland, 260, 303, 306, 308
Seaports (see Ports)
Selkirk, county and town, 357
Settle, Valley of, 158
Severn Basin, counties of the, 133, 134
Severn, river, 37, 46, 48, 78, 97, 100, 104, 107, 135, 139, 223, 226
Severn Valley, 135, 137
Shap Fell, 43, 47, 121
Shannon, river and lakes, 367—369
Sheerness, 193
Shelve Hill, 135
Sheppey, Isle of, 33, 191, 193
Sherborne, 203
Sherwood Forest, 153
Shetland Islands, 308, 309
Shetland, 318, 319
Shiffnal, 136
Ship-building in Scotland, 300 318
Shires, or Counties (see Counties)
Shoeburyness, 180
Shrewsbury, 136
Shropshire, 133, 134
Sidlaw Hills, 279, 282, 294
Silver in Cornwall, 213; in the Isle of Man, 243
Silvermine Hills, 366
Skiddaw, 43, 55, 120
Skye, Isle of, 309, 323, 324
Slaney, river, 370
Slate and Slate-quarries, 12, 120, 217, 218, 219, 220, 221

Sleat, Sound of, 309
Slievebloom Hills, 366
Slieve-Donard, 366
Slieve League, 364
Sligo Bay, 373
Sligo, county and town, 384
Snaefell, 243
Snowdon range of Mountains, 47. 219
Snowdon, 24, 48, 55, 219
Soar, river, 86
Social differences of Ireland and Great
    Britain, 8
Soil of the British Islands, 6
Solent, the, 34
Solway Firth, 20, 36, 43, 300, 307, 308
Solway Moss, 355
Somersetshire, 205—207
Sound of Islay, 310
Sound of Mull, 310
Sound of Sleat, 309
Sources of rivers, 79
Southern waterparting, Counties of
    the, 194
Southern Counties of England, 110,
    111
South-western counties of England,
    110, 111, 205, 206
Southampton, Plain of, 74
Southampton as a port, 74
Southampton and Southampton Water,
    35, 94, 199
South Downs, 26, 60, 66, 67, 196, 197
South Esk, 293
South Foreland, 34
South Shields, 119
South Uist, 310
Southwark, 190
Spalding, 166
Sperrin Mountains, 364
Spey, river and valley, 264, 290
Spithead, 34
Start Point, 34
Staffordshire coal-fields and manufac-
    tures, 151
Stamford, 166
Stirling, county and town, 341, 342
Stiper Stones, 49, 135
Stockton-on-Tees, 119
Stone and Stone quarries, 27, 42, 211,
    217, 219
Stonehaven, 332
Stonehenge, 200
Stormont, 314
Stornoway, 323
Stort, river, 179, 183
Stour, river, 93, 94, 98, 175, 179, 191,
    192, 203
Stourbridge, 138
Strabane, 381

Stranraer, 354
Straits of Dover, 26
Strangford Bay, 374
Stratford-on-Avon, 141
Strathmore, 264, 282
Straths of Scotland, 286
Straw-plaiting, 169, 183, 185
Stroud, 139
Suck, river, 369
Sudbury, 176
Suffolk, 172, 173, 175
Sugarloaf Mount, 143
Suir, river, 370
Sunderland, 118
Surrey, 179, 188
Sussex, 194—196
Sutherland, 318, 320
Swale, river, 84
Swale, East, river, 93
Swamps, 88, 180
Swansea Bay, 37
Swansea, 225, 240
Swindon, 201

T.

TAFF, river, 224, 227, 239
Tain, 322
Tamar, river, 50, 94, 210, 212
Tame, river, 86, 141, 150
Tara, Hill of, 391
Tarns of the Welsh mountainous dis-
    trict, 48
Taunton, Vale of, 49, 209
Taunton, 210
Tavistock, 211
Tavy, river, 211
Tawe, river, 95, 211, 212, 224, 227
Tay, river, 293, 305
Tees, river, 33, 80, 82, 107, 118, 156
Teify, river, 100, 223, 226
Teign, river, 211
Teignmouth, 212
Teith, river, 296
Teme, river, 98, 135, 137
Temperature of the British Islands, 4
Tern, river, 98, 135
Test, river, 94, 198
Teviot, river, 298
Teviotdale, 315
Tewkesbury, 139
Thame, river, 92; and valley, 184,
    185, 186
Thames, the river, 26, 33, 107, 181,
    182, 187; its upper valley, 71; its
    basin and estuary, 72, 80, 90—92,
    138

Thames Basin, Counties of the, 177;
    map of towns and rivers, 178
Thanet, Isle of, 191
Thetford, 175
Thirlmere, 45
Thurles, 389
Thurso, 320
Tides, 7
Tin mines, 52, 213
Tinto Hill, 280
Tipperary, county and town, 389, 390
Tiverton, 212
Tone, river, 209
Tor Bay, 222
Torquay, 212
Torridge, river, 95, 212
Towy, river, 100, 223, 227
Trade, effect of geography on, 40 (see
    Commerce)
Tralee, Bay of, 373
Tralee, 388
Trent, river, its valleys and tributaries,
    83, 86, 87, 107, 165
Trent, Counties of the basin of the,
    145—147
Trim, 391
Trossachs, the, 268, 335
Trostan, Mount, 364
Trowbridge, 202
Truro, 213
Tuam, 385
Tullamore, 392
Tullow, 396
Tummel, river, 294, 295
Tunbridge, 191
Tunbridge Wells, 191
Tweed, river, 20, 80, 297
Tweeddale, 315
Tyne, river, 80, 82, 107, 117, 118
Tynemouth, 117
Tyrone, county, 381

U.

ULLESWATER, 45, 121
Ulster, 377, 378
Undercliff, Isle of Wight, 200
Upland ranges, 57
Uplands of the East Midland Coun-
    ties, 165
Uplands, Chalk, 172, 183, 198
Uplands, Oolitic, 163, 185, 200, 202,
    203
Uppingham, 166
Ure, river and valley, 84
Usk, river, 78, 99, 100, 107, 142, 143,
    223, 226
Usk, 143

V.

VALE OF AYLESBURY, 71, 184
Vale of Blackmore, 202
Vale of Catmoss, 166
Vale of Evesham, 137
Vale of Glamorgan, 240
Vale of Llangollen, 231
Vale of Pewsey, 201
Vale of Pickering, 84
Vale of Taunton, 49, 209
Vale of White Horse, 71, 187
Valencia Island, 374
Valleys, 69
Valley of the Ouse, 71
Valley of the Thames, 72
Valleys of Worcestershire, 137
Ventnor, 200
Virnwy, river, 98, 235

W.

WALES: political geography, 11; com-
    pared with Scotland, 12; general
    view of Wales, 19; the Welsh moun-
    tains, 11, 24, 41, 45, 46, 49, 54;
    area and population, 18, 216; rivers,
    78, 96, 100
Wallingford, 188
Wantage, 188
Ware, 183
Wareham, 203
Warrington, 128
Warwickshire, 133, 140
Warwick, 141
Wash, the, 33
Wash, estuary and rivers of the, 88, 89
Wash, Counties of the, or East Mid-
    land Counties, 162, 164
Wash Water, 120
Wastwater, 45
Waterford, county and town, 389
Waterford Harbour, 375
Waterford Hills, 366
Waterparting of England and Wales,
    20, 21
Waterparting of Southern England,
    207
Waterparting of Wales, 223
Waterparting of Scotland, 255, 256,
    287, 352
Watershed of the Bristol Channel, 96,
    226, 228, 234
Watershed of the Irish Sea, 102, 227,
    228
Watershed of the North Sea, 81
Watershed of Scotland, western, 250,
    eastern, 261

Watford, 184
Waveney, river, 93, 174, 175
Weald, the, 73, 74
Weald, rivers and district of the, 93, 196
Wealden Heights, the, 73, 78, 188—190, 194, 196
Wear, river, 80, 82
Weaver, river and valley, 70, 103
Weaver Hills, 150
Welland, river, 89, 107, 165—167
Wellingborough, 168
Wellington, Shropshire, 136
Wellington, Somerset, 210
Wells, Norfolk, 174
Welsh mountains, rivers of the, 96 ; coal-fields of the, 113
Welshpool, 236
Wenlock Edge, 49, 135
Wensum, river, 174
Westbury, 202
Westmeath, county, 391
West Midland Counties of England, 110, 111, 133
Westmoreland, Moors of, 114
Westmoreland, county, 121
Westport, 385
Wexford Haven, 375
Wexford, county and town, 395
Wey, river, 92, 189
Weymouth, 203
Wharfe, river, and its valley, 84
Whernside, 41, 43
Whitchurch, 136
Whiteadder, river, 298
White Coomb, 274
Whitehaven, 121
White Horse, Vale of, 71
White Horse Hill, 65, 187
Whiten Head, 306
Wick, 320
Wicklow, county, 394
Wicklow Head, 375
Wicklow Mountains, 366, 374
Wigan, 127
Wight, Isle of, 199
Wigtonshire, county and town, 352, 354
Wigton, 120
Wigton Bay, 508

Wiley, river, 200
Wiltshire, 194, 195, 200
Winchester, 198
Windermere, 45, 121, 129
Windrush, river, 186
Winds, 4
Windsor, 188
Wisbeach, 171
Wishaw, 347
Wisp Hill, 275
Witham, river, 88, 107, 165
Witney, 186
Wolds of York and Lincoln, 83
Wolds of Lincolnshire, 165
Woodstock, 186
Woollen trade of Yorkshire, 42, 104
Woolwich, 192
Worcestershire, 133, 137
Worcester, 137
Worthing, 197
Wrekin, the, 135
Wrexham, 232
Wye, river, 48, 49, 78, 99, 100, 107, 142, 223, 226
Wyre, Forest of, 135
Wyre, river, 125

Y.

Yarmouth, 174
Yare, river, 93, 107, 174
Yarrow, river, 298
Yeo, river, 209
Yeovil, 209
Yes Tor, 50, 211
Yorkshire, manufactures of, 42
York, Plain of, 69, 155, 156
Yorkshire Moors, 62, 155, 158
Yorkshire Wolds, 61, 83, 155, 158
York, the capital of Northern England, 157
Youghal harbour, 370
Youghal, 388
Ystwith, river, 223, 226

Z.

Zinc, 232

LONDON : R. CLAY, SONS, AND TAYLOR, PRINTERS.

www.ingramcontent.com/pod-product-compliance
Lightning Source LLC
Chambersburg PA
CBHW021343210326
41599CB00011B/727